STUDENT POCKET GUIDE

S0-BNZ-805

BIMAN DAS
STATE UNIVERSITY OF NEW YORK, POTSDAM

COLLEGE PHYSICS

WILSON
BUFFA
LOU

SIXTH EDITION

PEARSON

Prentice
Hall

Upper Saddle River, NJ 07458

Associate Editor: Christian Botting
Project Manager: Kathleen Boothby Sestak
Senior Editor: Erik Fahlgren
Editor-in-Chief, Science: Dan Kaveney
Editorial Assistant: Jessica Berta
Executive Managing Editor: Kathleen Schiaparelli
Assistant Managing Editor: Karen Bosch
Production Editor: Gina M. Cheselka
Supplement Cover Manager: Paul Gourhan
Supplement Cover Designer: Christopher Kossa
Manufacturing Buyer: Ilene Kahn
Manufacturing Manager: Alexis Heydt-Long
Cover Image: Greg Epperson/Index Stock Imagery

© 2007 Pearson Education, Inc.
Pearson Prentice Hall
Pearson Education, Inc.
Upper Saddle River, NJ 07458

All rights reserved. No part of this book may be reproduced in any form or by any means, without permission in writing from the publisher.

Pearson Prentice Hall™ is a trademark of Pearson Education, Inc.

The author and publisher of this book have used their best efforts in preparing this book. These efforts include the development, research, and testing of the theories and programs to determine their effectiveness. The author and publisher make no warranty of any kind, expressed or implied, with regard to these programs or the documentation contained in this book. The author and publisher shall not be liable in any event for incidental or consequential damages in connection with, or arising out of, the furnishing, performance, or use of these programs.

This work is protected by United States copyright laws and is provided solely for teaching courses and assessing student learning. Dissemination or sale of any part of this work (including on the World Wide Web) will destroy the integrity of the work and is not permitted. The work and materials from it should never be made available except by instructors using the accompanying text in their classes. All recipients of this work are expected to abide by these restrictions and to honor the intended pedagogical purposes and the needs of other instructors who rely on these materials.

Printed in the United States of America

10 9 8 7 6 5 4 3 2 1

ISBN 0-13-149718-9

Pearson Education Ltd., *London*
Pearson Education Australia Pty. Ltd., *Sydney*
Pearson Education Singapore, Pte. Ltd.
Pearson Education North Asia Ltd., *Hong Kong*
Pearson Education Canada, Inc., *Toronto*
Pearson Educación de Mexico, S.A. de C.V.
Pearson Education—Japan, *Tokyo*
Pearson Education Malaysia, Pte. Ltd.

Dedicated to the memory of my father,
Mr. Manindranath Das

Preface

The *Student Pocket Guide* is written to accompany *College Physics* by J. D. Wilson, A. J. Buffa, and B. Lou. It contains basic concepts, principles, equations, and applications from each section of *College Physics* as well as hints and suggestions for solving problems in physics. This mini-text is written based on my several years of experience teaching introductory physics.

This small paperback mini-text is easy to carry to class for taking notes. Students will also be able to customize it to meet their needs by adding notes into the margins. Students will find the *Pocket Guide* useful for reviewing basic concepts of physics, solving homework assignments, and pre-exam review. When students use physics texts, they usually highlight the material that they think they will need for tests or exams. I have found that some students are better than others in highlighting such important material. Also students often find it difficult to pay attention to the class lecture while simultaneously taking notes. Students will be able to use this *Pocket Guide* as an alternative to highlighting material from the text. They will be able to pay more attention to the class lecture since all concepts have been summarized in this mini-text.

Each chapter in the *Student Pocket Guide* begins with chapter objectives and a list of the key terms and concepts that the students will learn from the chapter. Each section of the mini-text follows the same order as *College Physics,* and it summarizes important concepts from each section in an easy-to-understand language. Learning how to approach and solve real-world problems is a basic part in any physics. To master problem solving, I have given hints and suggestions for problem solving at the end of each chapter based on my experience of working closely with students. I have included the key concepts and equations students need to use for solving problems in different sections of each chapter. Wherever possible, I have pointed out common mistakes that students often make in solving physics problems to make them aware of these.

I would appreciate any feedback about the *Student Pocket Guide* from students, teachers, and any other user of the book. As an author, I am grateful to Prentice Hall for providing me with this opportunity to work on this mini-text. I thank Christian E. Botting, Associate Editor, and the Physics and Astronomy team of Prentice Hall for their constant support and cooperation throughout the entire process. I thank my wife Indrani Das and my daughters Debapria Das and Deea Das for their patience, constant support, and cooperation, without which I would not have found the time to write the book.

It was a rewarding experience for me to write the *Student Pocket Guide*. I hope students will enjoy studying it and find it useful.

Biman Das
Professor of Physics
SUNY, Potsdam
dasb@potsdam.edu

How to Use the
Student Pocket Guide

The primary objective of physics is to discover the fundamental principles that govern the universe. It is fun to be able to understand why the universe behaves the way it does. This *Student Pocket Guide* is a mini-text that is intended to help you understand the concepts, laws, principles, equations, and applications of physics, as outlined in *College Physics* by Wilson, Buffa, and Lou. Although this mini-text is a condensed version of *College Physics*, it is not meant to replace it. The text contains worked-out problems, practice problems, conceptual questions, examples, and other features that are not included in this mini-text. Use this mini-text as a reference guide and as a notebook. This small paperback mini-text is easy to carry when you go home. You can read it when you are on a bus, train, airplane, or even when in a restaurant!

I suggest the following ways to use the *Student Pocket Guide*:

> Read the relevant chapter from the *Pocket Guide* before you attend class lectures.

> Carry the book to class to take brief notes in the margins where necessary.

> Read carefully the *Hints and Suggestions for Solving Problems* at the end of the chapter before you practice problems. Be wary of the common mistakes pointed out in the hints and suggestions.

Use the important equations and concepts from the *Student Pocket Guide* to complete your homework assignments.

Review the chapters of the *Student Pocket Guide* before you take any quiz, test, or exam.

Use the equations and the concepts from the mini-text where necessary to analyze experimental data in your laboratory.

Use the book as a reference for important concepts, laws, equations, and applications even after you finish the course.

Review the concepts and important equations before you take other physics tests such as MCAT.

Biman Das
Professor of Physics
SUNY, Potsdam
dasb@potsdam.edu

Contents

CHAPTER 1
MEASUREMENTS AND PROBLEMS SOLVING 1
1.1   Why and How We Measure 1
1.2   SI Units of Length, Mass, and Time 2
1.3   More about the Metric System 4
1.4   Unit Analysis 5
1.5   Unit Conversions 6
1.6   Significant Figures 6
1.7   Problem Solving 8
        Hints and Suggestions for Solving Problems 9

CHAPTER 2
KINEMATICS: DESCRIPTION OF MOTION 10
2.1   Distance and Speed: Scalar Quantities 10
2.2   One-Dimensional Displacement and Velocity:
        Vector Quantities 11
2.3   Acceleration 14
2.4   Kinematic Equations (Constant Acceleration) 16
2.5   Free Fall 17
        Hints and Suggestions for Solving Problems 18

CHAPTER 3
MOTION IN TWO DIMENSIONS 20
3.1   Components of Motion 20
3.2   Vector Addition and Subtraction 22
3.3   Projectile Motion 27
3.4   Relative Velocity 29
        Hints and Suggestions for Solving Problems 31

CHAPTER 4
FORCE AND MOTION 33
4.1   The Concepts of Force and Net Force 33
4.2   Inertia and Newton's First Law of Motion 34
4.3   Newton's Second Law of Motion 34
4.4   Newton's Third Law of Motion 35
4.5   More on Newton's Laws: Free-Body Diagrams and Translational Equilibrium 37
4.6   Friction 39
      Hints and Suggestions for Solving Problems 42

CHAPTER 5
WORK AND ENERGY 44
5.1   Work Done by a Constant Force 44
5.2   Work Done by a Variable Force 46
5.3   The Work–Energy Theorem: Kinetic Energy 47
5.4   Potential Energy 48
5.5   The Conservation of Energy 50
5.6   Power 52
      Hints and Suggestions for Solving Problems 53

CHAPTER 6
LINEAR MOMENTUM AND COLLISIONS 55
6.1   Linear Momentum 55
6.2   Impulse 56
6.3   The Conservation of Linear Momentum 58
6.4   Elastic and Inelastic Collisions 58
6.5   Center of Mass 61
6.6   Jet Propulsion and Rockets 63
      Hints and Suggestions for Solving Problems 64

CHAPTER 7
CIRCULAR MOTION AND GRAVITATION 65

7.1   Angular Measure 65
7.2   Angular Speed and Velocity 67
7.3   Uniform Circular Motion and Centripetal
        Acceleration 69
7.4   Angular Acceleration 71
7.5   Newton's Law of Gravitation 71
7.6   Kepler's Laws and Earth Satellites 74
        Hints and Suggestions for Solving Problems 78

CHAPTER 8
ROTATIONAL MOTION AND EQUILIBRIUM 80

8.1   Rigid Bodies, Translations, and Rotations 80
8.2   Torque, Equilibrium, and Stability 81
8.3   Rotational Dynamics 84
8.4   Rotational Work and Kinetic Energy 85
8.5   Angular Momentum 87
        Hints and Suggestion for Solving Problems 89

CHAPTER 9
SOLIDS AND FLUIDS 91

9.1   Solids and Elastic Moduli 92
9.2   Fluids: Pressure and Pascal's Principle 96
9.3   Buoyancy and Archimedes' Principle 99
9.4   Fluid Dynamics and Bernoulli's Equation 101
*9.5   Surface tension, Viscosity, and Poiseuille's Law 103
        Hints and Suggestions for Solving Problems 105

CHAPTER 10
TEMPERATURE AND KINETIC THEORY 107
10.1 Temperature and Heat 107
10.2 The Celsius and Fahrenheit Temperature Scales 108
10.3 Gas Laws, Absolute Temperature, and the
Kelvin Temperature Scale 110
10.4 Thermal Expansion 113
10.5 The Kinetic Theory of Gases 115
*10.6 Kinetic Theory, Diatomic Gases, and the
Equipartition Theorem 117
Hints and Suggestions for Solving Problems 119

CHAPTER 11
HEAT 121
11.1 Definition and Units of Heat 121
11.2 Specific Heat and Calorimetry 122
11.3 Phase Changes and Latent Heat 124
11.4 Heat Transfer 127
Hints and Suggestions for Solving Problems 130

CHAPTER 12
THERMODYNAMICS 132
12.1 Thermodynamic Systems, States, and Processes 133
12.2 The First Law of Thermodynamics 134
12.3 Thermodynamic Processes for an Ideal Gas 135
12.4 The Second Law of Thermodynamics
and Entropy 137
12.5 Heat Engines and Thermal Pumps 139
12.6 The Carnot Cycle and Ideal Heat Engines 142
Hints and Suggestions for Solving Problems 144

CHAPTER 13
VIBRATIONS AND WAVES 146
13.1   Simple Harmonic Motion 146
13.2   Equations of Motion 149
13.3   Wave Motion 152
13.4   Wave Properties 154
13.5   Standing Waves and Resonance 156
        Hints and Suggestions for Solving Problems 159

CHAPTER 14
SOUND 160
14.1   Sound waves 160
14.2   The Speed of Sound 162
14.3   Sound Intensity and Sound Intensity Level 162
14.4   Sound Phenomena 164
14.5   The Doppler Effect 165
14.6   Musical Instruments and Sound Characteristics 169
        Hints and Suggestions for Solving Problems 171

CHAPTER 15
ELECTRIC CHARGE, FORCES, AND FIELDS 173
15.1   Electric Charge 173
15.2   Electrostatic Charging 174
15.3   Electric Force 177
15.4   Electric Field 177
15.5   Conductors and Electric Fields 180
*15.6   Gauss's Law for Electric Fields: A Qualitative
        Approach 181
        Hints and Suggestions for Solving Problems 182

CHAPTER 16
ELECTRIC POTENTIAL, ENERGY, AND CAPACITANCE 184

16.1   Electric Potential Energy and Electric Potential Difference 184
16.2   Equipotential Surfaces and the Electric Field 188
16.3   Capacitance 189
16.4   Dielectrics 191
16.5   Capacitors in Series and in Parallel 193
        Hints and Suggestions for Solving Problems 194

CHAPTER 17
ELECTRIC CURRENT AND RESISTANCE 196

17.1   Batteries and Direct Current 196
17.2   Current and Drift Velocity 198
17.3   Resistance and Ohm's Law 200
17.4   Electric Power 203
        Hints and Suggestions for Solving Problems 205

CHAPTER 18
BASIC ELECTRIC CIRCUITS 206

18.1   Resistances in Series, Parallel, and Series-Parallel Combinations 206
18.2   Multiloop Circuits and Kirchhoff's Rules 209
18.3   *RC* Circuits 211
18.4   Ammeters and Voltmeters 213
18.5   Household Circuits and Electrical Safety 214
        Hints and Suggestions for Solving Problems 216

CHAPTER 19
MAGNETISM 218
19.1   Magnets, Magnetic Poles, and Magnetic Field
        Direction 219
19.2   Magnetic Field Strength and Magnetic Force 220
19.3   Applications: Charged Particles in
        Magnetic Fields 222
19.4   Magnetic Forces on Current-Carrying Wires 224
19.5   Applications: Current-Carrying Wires in Magnetic
        Fields 226
19.6   Electromagnetism: The Source of Magnetic
        Fields 227
19.7   Magnetic Materials 230
*19.8 Geomagnetism: Earth's Magnetic Field 232
        Hints and Suggestions for Solving Problems 234

CHAPTER 20
ELECTROMAGNETIC INDUCTION AND
WAVES 236
20.1   Induced Emf's: Faraday's Law and Lenz's Law 237
20.2   Electric Generators and Back Emf 240
20.3   Transformers and Power Transmission 242
20.4   Electromagnetic Waves 244
        Hints and Suggestions for Solving Problems 248

CHAPTER 21
AC CIRCUITS 249
21.1   Resistance in an AC Circuit 250
21.2   Capacitive Reactance 252
21.3   Inductive Reactance 254
21.4   Impedance: *RLC* Circuits 255
21.5   Circuit Resonance 259
        Hints and Suggestions for Solving Problems 261

CHAPTER 22
REFLECTION AND REFRACTION OF LIGHT 263
22.1   Wave Fronts and Rays 263
22.2   Reflection 264
22.3   Refraction 265
22.4   Total Internal Reflection and Fiber Optics 267
22.5   Dispersion 269
        Hints and Suggestions for Solving Problems 270

CHAPTER 23
MIRRORS AND LENSES 271
23.1   Plane Mirrors 271
23.2   Spherical Mirrors 273
23.3   Lenses 277
23.4   The Lens Maker's Equation 280
*23.5   Lens Aberrations 281
        Hints and Suggestions for Solving Problems 282

CHAPTER 24
PHYSICAL OPTICS: THE WAVE NATURE OF LIGHT 284
24.1   Young's Double-Slit Experiment 285
24.2   Thin-Film Interference 287
24.3   Diffraction 289
24.4   Polarization 292
*24.5   Atmospheric Scattering of Light 296
        Hints and Suggestions for Solving Problems 297

CHAPTER 25
VISION AND OPTICAL INSTRUMENTS 299
25.1   The Human Eye 299
25.2   Microscopes 302
25.3   Telescopes 304
25.4   Diffraction and Resolution 305
*25.5   Color 307
        Hints and Suggestions for Solving Problems 308

CHAPTER 26
RELATIVITY 309
26.1   Classical Relativity and the Michelson–Morley
        Experiment 310
26.2   The Postulates of Special Relativity and the
        Relativity of Simultaneity 311
26.3   The Relativity of Length and Time: Time Dilation
        and Length Contraction 312
26.4   Relativistic Kinetic Energy, Momentum, Total
        Energy, and Mass–Energy Equivalence 315
26.5   The General Theory of Relativity 317
*26.6   Relativistic Velocity Addition 319
        Hints and Suggestions for Solving Problems 320

CHAPTER 27
QUANTUM PHYSICS 322
27.1   Quantization: Planck's Hypothesis 323
27.2   Quanta of Light: Photons and the Photoelectric
        Effect 325
27.3   Quantum "Particles": The Compton Effect 327
27.4   The Bohr Theory of the Hydrogen Atom 329
27.5   A Quantum Success: The Laser 333
        Hints and Suggestions for Solving Problems 336

CHAPTER 28
QUANTUM MECHANICS AND ATOMIC PHYSICS 339
28.1   Matter Waves: The De Broglie Hypothesis 339
28.2   The Schrödinger Wave Equation 342
28.3   Atomic Quantum Numbers and the Periodic Table 343
28.4   The Heisenberg Uncertainty Principle 346
28.5   Particles and Antiparticles 347
        Hints and Suggestions for Solving Problems 349

CHAPTER 29
THE NUCLEUS 351
29.1   Nuclear Structure and the Nuclear Force 352
29.2   Radioactivity 354
29.3   Decay Rate and Half-Life 359
29.4   Nuclear Stability and Binding Energy 361
29.5   Radiation Detection and Applications 363
        Hints and Suggestions for Solving Problems 367

CHAPTER 30
NUCLEAR REACTIONS AND ELEMENTARY PARTICLES 369
30.1   Nuclear Reactions 369
30.2   Nuclear Fission 371
30.3   Nuclear Fusion 374
30.4   Beta Decay and the Neutrino 376
30.5   Fundamental Forces and Exchange Particles 377
30.6   Elementary Particles 380
30.7   The Quark Model 381
30.8   Force Unification Theories, the Standard Model, and the Early Universe 382
        Hints and Suggestions for Solving Problems 384

CHAPTER 1
MEASUREMENT AND
PROBLEM SOLVING

Measurement is an integral part of physics. This chapter discusses physical quantities and the different systems of units used in measurements. Unit analysis, significant figures, and problem solving strategy are also discussed.

Key Terms and Concepts

> Standard units
> Length, mass, and time
> Internal system of units
> Base and derived units
> Meter, kilogram, and second
> Mks, cgs, and fps systems
> Prefix
> Volume and density
> Significant figures
> Unit conversions
> Unit analysis
> Order of magnitude
> Problem solving

1.1 Why and How We Measure

Physics attempts to describe nature and to discover the fundamental principles governing nature. Measurement is one of the important tools used to describe nature and discover its fundamental principles.

1

2007 Pearson Education. Inc., Upper Saddle River, NJ. All rights reserved. This material is protected under all copyright laws as they currently exist. No portion of this material may be reproduced, in any form or by any means, without permission in writing from the publisher.

Standard Units

Measurements are expressed in units, such as the *foot*, which is used for length measurements. An officially accepted unit is called a **standard unit**. A group or combination of standard units is called a **system of units**.

1.2 SI Units of Length, Mass, and Time

The fundamental physical quantities are length, mass, and time. The system of units adopted by scientists to represent physical quantities is based on the metric system. The modernized version of the metric system is known as the **International System of Units** (SI). The SI includes *base quantities* and *derived quantities*, which are described by base units (such as meter) and derived units (such as meter per second), respectively.

Length

Length is a base quantity that determines the separation between two points. The SI unit of length is the **meter** (m). It was originally defined as 1/10 000 000 of the distance from the North Pole to the equator along a meridian passing through Paris. In 1983, the meter was redefined to be the distance traveled by light in a vacuum in 1/299 792 458 of a second.

Mass

Mass is a base quantity. It describes the amount of matter in a substance. The SI unit of mass is the **kilogram (kg)**. One kilogram is the mass of a specially designed platinum-iridium alloy cylinder at the International Bureau of Weights and Standards in Sevres, France. An object has the same mass regardless of its location. This is not true for its weight. Keep in mind that *weight* is proportional to the *mass*, but not the same as the mass.

2

© 2007 Pearson Education, Inc., Upper Saddle River, NJ. All rights reserved. This material is protected under all copyright laws as they currently exist. No portion of this material may be reproduced, in any form or by any means, without permission in writing from the publisher.

Time

Time is a base quantity. It represents the forward flow of events or the interval between two events. Time is said to be a fourth dimension. The SI unit of time is the **second (s)**. A second was defined as 1/86 400 of a mean solar day. In 1967, an atomic standard was adopted as a more precise definition of a second. One second is defined by the radiation frequency of a cesium-133 atom. An atomic clock maintains the time standard with a variation of about 1 second in 300 years. In 1999, another cesium-133 clock, called the atomic fountain clock, was adopted that maintains time standard with a variation of less than 1 second in 20 million years. An even more precise clock, called the all-optical atomic clock, is under development. It uses a single cooled ion of liquid mercury linked to a laser oscillator. The frequency of mercury atoms is 100 000 times higher than that of cesium atoms, and thus it can measure time intervals that are 100 000 times shorter than those of cesium atomic clocks.

SI Base Units

The complete SI has seven base quantities and base units as shown in the following table.

Name of Unit	Property Measured
meter (m)	length
kilogram (kg)	mass
second (s)	time
ampere (A)	electric current
kelvin (K)	temperature
mole (mol)	amount of substance
candela (cd)	luminous intensity

3

2007 Pearson Education, Inc., Upper Saddle River, NJ. All rights reserved. This material is protected under all copyright laws as they currently exist. No portion of this material may be reproduced, in any form or by any means, without permission in writing from the publisher.

1.3 More about the Metric System

The metric system involving the standard units of length, mass, and time used to be called the **mks system** (for meter-kilogram-second). Another metric system sometimes used for relatively small quantities is called the **cgs system**. The British engineering (BE) system is often encountered in everyday use in the United States. The British system is sometimes called the **fps system** (foot-pound-second). The metric system is used throughout the world and is in increasing use in the United States.

Standard **prefixes** are used to designate common multiples in powers of ten. For example:

$$1 \text{ kilogram (kg)} = 10^3 \text{ gram (g)}$$
$$1 \text{ megawatt (MW)} = 10^6 \text{ watt (W)}$$
$$1 \text{ gigahertz (GHz)} = 10^9 \text{ Hertz (Hz)}$$
$$1 \text{ millimeter (mm)} = 10^{-3} \text{ meter (m)}$$
$$1 \text{ micrometer } (\mu\text{m}) = 10^{-6} \text{ meter (m)}$$
$$1 \text{ nanometer (nm)} = 10^{-9} \text{ meter (m)}$$
$$1 \text{ picosecond (ps)} = 10^{-12} \text{ second (s)}$$
$$1 \text{ femtometer (fm)} = 10^{-15} \text{ meter (m)}$$

Nanotechnology involves technology on the nanometer scale. It involves the manufacture or building of things one atom or molecule at a time, so nanometer is the scale. Nanotechnology presents the possibility of constructing novel molecular devices or "machines" with extraordinary properties and abilities. The potential applications include medical delivery (delivering a drug directly to a particular site, such as a cancerous growth), nanocomputing (making smaller and faster computing chips), and fabric improvement (a nanostructure that can be attached to fibers).

4

© 2007 Pearson Education, Inc., Upper Saddle River, NJ. All rights reserved. This material is protected under all copyright laws as they currently exist. No portion of this material may be reproduced, in any form or by any means, without permission in writing from the publisher.

Volume

The volume of a substance is the amount of space occupied by the substance. The SI unit of volume is the cubic meter (m^3). A common nonstandard unit of volume is **liter (L)**. $1 \text{ L} = 1000 \text{ cm}^3$.

1.4 Unit Analysis

The fundamental or base quantities used in physical descriptions are called *dimensions*. For example, length, mass, and time are dimensions. The distance between two points can be measured in feet or meters etc., but the quantity will always have the dimension of length. Addition and subtraction of two or more quantities is meaningful only when the quantities have the same dimension.

Dimensions provide a procedure by which consistency of an equation can be checked. Using units to check a equation is called **unit analysis**. For example, in the equation $x = at^2$ (where x, a, and t represent position acceleration, and time, respectively), $m = (m/s^2)(s^2) = m$. Thus, the equation is dimensionally correct. In an equation each term must have the same units on both sides. In the derivation of formulas, unit checks help verify each step of the derivation. If an equation is correct by unit analysis, it must be dimensionally correct. However, a dimension check is not a guarantee that the formula is correct because dimensionless factors (such as ½) do not show up in the dimension check.

5

© 2007 Pearson Education, Inc., Upper Saddle River, NJ. All rights reserved. This material is protected under all copyright laws as they currently exist. No portion of this material may be reproduced, in any form or by any means, without permission in writing from the publisher.

Mixed Units
In general, always use the same unit for a given dimension throughout the problem. The terms should not be added or subtracted without first converting them to the same units.

Determining the Units of Quantities
The unit of quantities can be obtained from their definitions. For example, density, ρ, is defined as

$$\rho = \frac{m}{V} \tag{1.1}$$

where m is the mass and V is the volume. Since the SI units of mass and volume are kg and m^3, respectively, the SI unit for density is kg/m^3.

1.5 Unit Conversions

Units can be converted from one set of units to another by multiplying by the appropriate factors, called **conversion factors.** Conversion factors are simply equivalence statements expressed in the form of ratios. Some important conversions are listed here:

$$1 \text{ mi} = 1.609 \text{ km}$$
$$1 \text{ ft} = 0.3048 \text{ m}$$
$$1 \text{ m} = 3.281 \text{ ft}$$
$$1 \text{ in.} = 0.0254 \text{ m}$$
$$1 \text{ rad} = 57.3°$$
$$1 \text{ lb} = 4.448 \text{ N}$$
$$1 \text{ km/hr} = 0.6214 \text{ mi/hr}$$

1.6 Significant Figures

The data given in a problem are either exact numbers or measured numbers. **Exact numbers** do not have any

6

© 2007 Pearson Education, Inc., Upper Saddle River, NJ. All rights reserved. This material is protected under all copyright laws as they currently exist. No portion of this material may be reproduced, in any form or by any means, without permission in writing from the publisher.

uncertainty or error. **Measured numbers** are obtained from experimental measurements and usually have some degree of uncertainty or error. The number of **significant figures** in a physical quantity is equal to the number of digits that are known with certainty.

Use the following rule to determine the significant figures of a quantity:

- Zeros at the beginning of a number are not significant.

- Zeros within a number are significant.

- Zeros at the end of a number after the decimal point are significant.

- In whole numbers without a decimal point that end in one or more zeros (for example, 500 kg), the zeros may or may not be significant. The ambiguity can be removed by using scientific notation:

  5.0×10^2 kg   two significant figures
  5.00×10^2 kg   three significant figures

The result of mathematical operations should be reported with the proper number of significant figures. Use the following rule to determine the number of significant figures:

1. When multiplying and dividing quantities, leave as many significant figures in the answer as there are in the quantity with the fewest number of significant figures.

7

© 2007 Pearson Education, Inc., Upper Saddle River, NJ. All rights reserved. This material is protected under all copyright laws as they currently exist. No portion of this material may be reproduced, in any form or by any means, without permission in writing from the publisher.

2. When adding or subtracting quantities, leave the same number of decimal places in the answer as there are in the quantity with the least number of decimal places.

Use the following rules to round off numbers:

1. If the first digit to be dropped is less than 5, leave the preceding digit as is (for example, 2.132 becomes 2.13 rounded to three significant figures).

2. If the first digit to be dropped is 5 or greater, increase the preceding digit by one (for example, 2.136 becomes 2.14 rounded to three significant figures).

1.7 Problem Solving

To learn physics, you need to practice by solving problems. Although there is no fixed recipe for problem solving, a few general points are worth keeping in mind:

- Read the problem carefully and analyze the probelm.
- Draw a sketch, if possible.
- List data and unknowns. Convert units, if necessary.
- Determine which principle(s) and equation(s) are applicable and how these can be applied.
- Substitute the given data into equatons and perform calculations to get the result.
- Consider whether the result makes sense.

Approximation and Order-of-Magnitude Calculations
Order-of-magnitude means that we express a quantity to the power of 10 closest to the actual value. For example, if we get the order of magnitude of time as 10^5 s, we expect

8

© 2007 Pearson Education, Inc., Upper Saddle River, NJ. All rights reserved. This material is protected under all copyright laws as they currently exist. No portion of this material may be reproduced, in any form or by any means, without permission in writing from the publisher.

that the exact number lies between 1×10^5 s and 10×10^5 s. Order-of-magnitude calculation is important when the exact solution is not necessary or cannot be done easily.

Hints and Suggestions for Solving Problems

1. Recognize the prefixes (such as k for kilo, M for mega, etc.), short notations (such as m for meter, s for seconds, etc.), and other SI symbols in a problem.

2. Make sure that all units are consistent in a given problem. Use conversion factors, if necessary, to make them consistent. Never combine different sets of units (such as millimeter, meter, and kilometer) in the same problem. Use SI units whenever possible.

3. Use unit analysis to check that the units of a physical quantity are correct. Make sure that the dimensions and units of each term in an equation are the same.

4. Express the result using scientific notation. Keep only the desired number of significant figures in the final answer to a problem.

5. Follow the strategy outlined in Section 1.7 when solving a problem.

9

2007 Pearson Education, Inc., Upper Saddle River, NJ. All rights reserved. This material is protected under all copyright laws as they currently exist. No portion of this material may be reproduced, in any form or by any means, without permission in writing from the publisher.

CHAPTER 2
KINEMATICS: DESCRIPTION OF MOTION

The branch of physics that is concerned with the study of motion with reference to its cause and effect is called mechanics. Mechanics is divided into two parts: (1) kinematics, and (2) dynamics. **Kinematics** describes the motion of objects without any regard for how the motion is caused. **Dynamics** analyzes the causes of motion. This chapter discusses kinematics for one-dimensional motion.

Key Terms and Concepts

> Distance
> Speed
> Average speed and instantaneous speed
> Displacement
> Velocity
> Average velocity and instantaneous velocity
> Acceleration
> Average acceleration and instantaneous acceleration
> Free fall
> Acceleration due to gravity

2.1 Distance and Speed: Scalar Quantities

Distance
The **distance** traveled by a moving object is the total path length traversed in moving from one location to another. In SI units, distance is expressed in meters (m). Distance is a scalar quantity. A **scalar quantity** is one that has only magnitude or size.

© 2007 Pearson Education, Inc., Upper Saddle River, NJ. All rights reserved. This material is protected under all copyright laws as they currently exist. No portion of this material may be reproduced, in any form or by any means, without permission in writing from the publisher.

Speed

For a moving object its distance changes with time. The time rate at which the distance of a moving object changes is called its **speed**. The **average speed** of a moving object is defined as:

$$\text{average speed} = \frac{\text{distance traveled}}{\text{total time to travel the distance}} \quad (2.1)$$

$$\bar{s} = \frac{d}{\Delta t} = \frac{d}{t_2 - t_1}$$

where d is the distance traveled and Δt is the total time elapsed in traveling that distance. The SI unit for speed is meters per second (m/s). The British standard unit of speed is feet/second (ft/s), but often it is expressed in miles/hour (mi/h). Since distance is a scalar quantity, speed is also a scalar.

Average speed gives the general description of motion over a time interval. If the time interval Δt becomes smaller and smaller and approaches zero, the speed calculation gives an **instantaneous speed**. Instantaneous speed tells how fast an object moves *at a particular instant of time*.

2.2 One-Dimensional Displacement and Velocity: Vector Quantities

Displacement

The **displacement** of a moving object is the straight-line distance between its final position and its initial position along with the *direction* from the starting position to the final position. Displacement is a **vector** quantity. A vector

11

2007 Pearson Education, Inc., Upper Saddle River, NJ. All rights reserved. This material is protected under all copyright laws as they currently exist. No portion of this material may be reproduced, in any form or by any means, without permission in writing from the publisher.

(short for vector quantity) has both magnitude *and* direction. Graphically, vectors are represented by arrows, with the direction of the arrowhead providing the direction of the vector and the length of the arrow proportional to the magnitude of the vector. In the text, vector quantities are indicated by boldface type.

$$x_1 \qquad \Delta x \qquad x_2$$

If the initial and final positions of an object are x_1 and x_2 at time t_1 and t_2, respectively, the displacement is

$$\Delta x = x_2 - x_1 \qquad (2.2)$$

In one dimension, there are only two possible directions of motion (left or right). The x-axis is commonly used for horizontal motions, and a plus (+) sign is used to indicate the direction to the right and a minus (−) sign is used to indicate the direction to the left. Displacement is expressed in meters, and it can be positive or negative.

Velocity
Velocity of an object tells how fast the object is moving *and* in what direction. The average velocity is the displacement divided by the total travel time:

$$\text{average velocity} = \frac{\text{displacement}}{\text{total travel time}} \qquad (2.3)$$

$$\bar{v} = \frac{\Delta x}{\Delta t} = \frac{x_2 - x_1}{t_2 - t_1}$$

The SI unit of velocity is meters per second (m/s).

12

© 2007 Pearson Education, Inc., Upper Saddle River, NJ. All rights reserved. This material is protected under all copyright laws as they currently exist. No portion of this material may be reproduced, in any form or by any means, without permission in writing from the publisher.

If all the motion is in one direction, the distance is equal to the magnitude of displacement, and the average speed is the magnitude of the average velocity. However, this is not true if the motion is in both directions.

As with speed, when the time interval approaches zero, we obtain **instantaneous velocity**. Mathematically,

$$v = \lim_{\Delta t \to 0} \frac{\Delta x}{\Delta t} \qquad (2.4)$$

The instantaneous velocity of an object tells how fast an object moves and in what direction at a particular instant of time. **Uniform motion** in one dimension means motion with a constant velocity.

Graphical Analysis

Graphical analysis is often helpful in understanding a motion. For example, the motion of a car may be represented on a plot of position versus time, or x versus t. A straight line is obtained on such a graph if the motion is uniform. The slope of the line, $\Delta x/\Delta t$, is equal to the average velocity. For uniform motion, average velocity is the same as the instantaneous velocity.

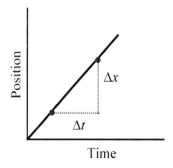

13

2007 Pearson Education, Inc., Upper Saddle River, NJ. All rights reserved. This material is protected under all copyright laws as they currently exist. No portion of this material may be reproduced, in any form or by any means, without permission in writing from the publisher.

For nonuniform velocities, the x-versus-t plot is a curved line, as shown below. The slope of the line between two positions is the average velocity between those positions, and the instantaneous velocity is the slope of the line tangent to the curve at any point. Instantaneous velocity can be positive, negative, or zero. Its sign indicates the direction of motion.

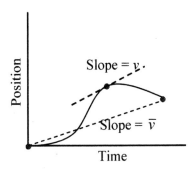

2.3 Acceleration

The time rate of change of velocity of an object is called its **acceleration**. The **average acceleration** is defined as the change in velocity divided by the time taken to make the change:

$$\text{average acceleration} = \frac{\text{change in velocity}}{\text{time to make the change}} \quad (2.5)$$

$$\bar{a} = \frac{\Delta v}{\Delta t} = \frac{v_2 - v_1}{t_2 - t_1} = \frac{v - v_0}{t - t_0}$$

The SI unit of velocity is meters per second per second (m/s^2).

14

© 2007 Pearson Education, Inc., Upper Saddle River, NJ. All rights reserved. This material is protected under all copyright laws as they currently exist.
No portion of this material may be reproduced, in any form or by any means, without permission in writing from the publisher.

For the special case of a linear motion, plus and minus signs can be used for the velocity directions, then

$$\bar{a} = \frac{v - v_0}{t} \qquad (2.6)$$

where t_0 is taken to be zero.

The acceleration at a particular time is called **instantaneous acceleration**. Mathematically,

$$a = \lim_{\Delta t \to 0} \frac{\Delta v}{\Delta t} \qquad (2.7)$$

Constant Acceleration

For motions with a constant acceleration, a, Equation (2.6) becomes,

$$v = v_0 + at \qquad (2.8)$$

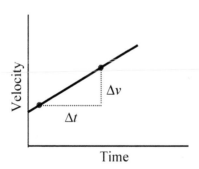

The acceleration of a moving object can be obtained from its velocity-versus-time graph. For motion with a constant acceleration, the velocity-versus-time plot is a straight line, and the slope of the line is the acceleration. If the slope is positive, the velocity of the object is increasing. If the slope is negative, the velocity of the object is decreasing (deceleration).

15

© 2007 Pearson Education, Inc., Upper Saddle River, NJ. All rights reserved. This material is protected under all copyright laws as they currently exist. No portion of this material may be reproduced, in any form or by any means, without permission in writing from the publisher.

When acceleration is constant, $\overline{v}$ is the average of the initial and final velocities:

$$\overline{v} = \frac{v + v_0}{2} \tag{2.9}$$

When the velocity and acceleration have the same sign, the speed of an object is increasing. When the velocity and acceleration have opposite signs, the speed of an object is decreasing.

2.4 Kinematic Equations (Constant Acceleration)

The equations describing the motion of objects with constant acceleration can be used to describe a wide range of everyday phenomena, such as the acceleration of a thrown ball. Description of motion with one dimension with constant acceleration requires only three basic equations, as given below:

$$x = x_0 + \overline{v} t \tag{2.3}$$

$$\overline{v} = \frac{v + v_0}{2} \tag{2.9}$$

$$v = v_0 + at \tag{2.8}$$

Also, from equations 2.3 and 2.9

$$x = x_0 + \frac{1}{2}(v + v_0)t \tag{2.10}$$

The following two equations can be derived from the preceding equations and are very useful to avoid multiple

16

© 2007 Pearson Education, Inc., Upper Saddle River, NJ. All rights reserved. This material is protected under all copyright laws as they currently exist. No portion of this material may be reproduced, in any form or by any means, without permission in writing from the publisher.

applications of the preceding equations in solving problems.

$$x = x_0 + v_0 t + \frac{1}{2} a t^2 \qquad (2.11)$$

$$v^2 = v_0^2 + 2a (x - x_0) \qquad (2.12)$$

Graphical Analysis of Kinematic Equations
When the acceleration of an object is constant, the plot of the velocity-versus-time graph is a straight line. The area under the plot is equal to the distance covered in the motion.

2.5 Free Fall

Objects in motion under the influence of gravity only are in free fall. If a ball is dropped, it is in free fall. Also, if a ball is thrown upward or downward, it is in free fall as soon as it is released.

Galileo first showed that all freely falling objects move with the same constant acceleration. Astronaut David Scott dropped a feather and a hammer on the moon in 1971 and found that both of them reached the lunar surface at the same time. The free-fall acceleration produced by the Earth's gravity is denoted by the symbol g and is called the **acceleration due to gravity**.

The value of g varies slightly from place to place on the surface of the Earth; near the surface of the Earth g is found to be about $+9.80$ m/s^2. Note that g is always positive. It is customary to use y to represent vertical direction and to take upward as positive. Thus, the acceleration of an object in free fall is $a = -g = -9.80$ m/s^2.

17

2007 Pearson Education, Inc., Upper Saddle River, NJ. All rights reserved. This material is protected under all copyright laws as they currently exist. No portion of this material may be reproduced, in any form or by any means, without permission in writing from the publisher.

Since g is approximately constant, the following kinematic equations can be used for the motion of an object in free fall:

$$v = v_0 - gt \qquad (2.8')$$

$$y = y_0 + v_0 t - \frac{1}{2}gt^2 \qquad (2.11')$$

$$v^2 = v_0^2 - 2g(y - y_0) \qquad (2.12')$$

$$y = y_0 + \frac{1}{2}(v + v_0)t \qquad (2.10')$$

Hints and Suggestions for Solving Problems

1. When solving a problem, read the problem several times. Determine the subcategory to which the problem belongs, such as motion with constant velocity or motion with constant acceleration.

2. Draw a sketch of the problem whenever possible.

3. Recognize the symbols Δx, Δv, and Δt that represent the displacement, velocity change, and time interval, respectively.

4. Be sure to note the difference between the two positions of an object to calculate its displacement. Also, remember to subtract the two times to find the time interval.

5. The slope of a position-versus-time graph is velocity. If the graph is linear, slope is a constant, and that represents motion with a constant velocity. Also, the slope of a velocity-versus-time graph is acceleration, and a linear velocity-versus-time graph represents

18

© 2007 Pearson Education, Inc., Upper Saddle River, NJ. All rights reserved. This material is protected under all copyright laws as they currently exist. No portion of this material may be reproduced, in any form or by any means, without permission in writing from the publisher.

motion with a constant acceleration. These concepts will be needed to solve some problems in Sections 2.2 and 2.3.

6. Instantaneous velocity and average velocity are the same if there is no acceleration or deceleration.

7. Remember that a negative value of acceleration does not necessarily mean a deceleration. An object may be accelerating, but its acceleration can have a negative sign. Consider velocity and acceleration to be positive if they are directed to the right and stick with the direction for the whole problem.

8. For problems involving motion with constant acceleration (such as those in Section 2.4), use the equations introduced in Section 2.4. Each of these equations relates four variables, and usually all of them will be given except one. Find the right equation to determine the unknown. Make a table like the following to enter data and unknowns:

v	v_0	a	t
Given	Given	Given	?

9. Free-fall motion means motion under the influence of gravity only. Problems involving free-fall motion (such as those in Section 2.5) can be solved using the equations from Section 2.5. In this case, $a = -g = -9.80 \text{ m/s}^2$. The negative sign reflects the fact that acceleration due to gravity is always directed downward.

19

©2007 Pearson Education, Inc., Upper Saddle River, NJ. All rights reserved. This material is protected under all copyright laws as they currently exist. No portion of this material may be reproduced, in any form or by any means, without permission in writing from the publisher.

CHAPTER 3
MOTION IN TWO DIMENSIONS

This chapter explores the study of kinematics of motion in two dimensions, particularly projectile motion. The rules of addition and subtraction of vectors and relative velocity are discussed.

Key Terms and Concepts

Components of motion
Vector addition and subtraction
Triangle method
Vector components
Analytical component method
Unit vectors
Relative velocity
Projectile motion
Time of flight
Range

3.1 Components of Motion

In Chapter 2 an object moving in a straight line was considered to be moving along either x or y axes. For the motion of an object along neither x or y axis (such as the rolling motion of a ball along the diagonal across a table with x and y axes as the edges of the table), both x and y coordinates are needed to describe the motion. We call this motion to be a motion in two dimensions. If the object has a velocity $\vec{v}$ in a direction at an angle θ with respect to the x axis, then the velocities in the x and y directions can be obtained by resolving the velocity vector into the **components of motion** in these directions. The velocity

20

© 2007 Pearson Education, Inc., Upper Saddle River, NJ. All rights reserved. This material is protected under all copyright laws as they currently exist. No portion of this material may be reproduced, in any form or by any means, without permission in writing from the publisher.

components will have the following magnitudes, respectively:

$$v_x = v \cos \theta \qquad (3.1a)$$
$$v_y = v \sin \theta \qquad (3.1b)$$

Here v_x and v_y are the x and y components of the velocity of the object, respectively. Here v is a combination of velocities in the x and y directions and is given by

$$v = \sqrt{v_x^2 + v_y^2}$$

We can determine the distance of the object in the x and y directions separately as follows:

$$x = x_0 + v_x t \qquad (3.2a)$$
$$y = y_0 + v_y t \qquad (3.2b)$$

The distance of the object from the origin is

$$d = \sqrt{x^2 + y^2}$$

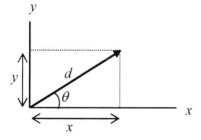

Kinematic Equations for Components of Motion

For a two-dimensional motion in a plane with constant velocity, the motion is in a straight line. For a motion in a plane with a constant acceleration having components a_x

21

2007 Pearson Education, Inc., Upper Saddle River, NJ. All rights reserved. This material is protected under all copyright laws as they currently exist. No portion of this material may be reproduced, in any form or by any means, without permission in writing from the publisher.

and a_y, the displacement and velocity components are given as in the kinematic equations of Chapter 2 in the x and y directions.

For an object starting with an initial velocity $\vec{v}_0$ and acceleration $\vec{a}$, the equations of motion for the x direction are:

$$x = x_0 + v_{x0}\,t + \frac{1}{2}\,a_x t^2 \qquad (3.3a)$$

$$y = y_0 + v_{y0}\,t + \frac{1}{2}\,a_y t^2 \qquad (3.3b)$$

$$v_x = v_{x0} + a_x t \qquad (3.3c)$$

$$v_y = v_{y0} + a_y t \qquad (3.3d)$$

If an object initially moving with a constant velocity suddenly experiences an acceleration in the direction of the velocity ($0°$) or opposite to it ($180°$), it would continue in a straight-line path either speeding up or slowing down, respectively. But, if the acceleration vector is at an angle different from $0°$ or $180°$ to the velocity vector, the motion will be along a curved path.

3.2 Vector Addition and Subtraction

Vector quantities such as displacement, velocity, and acceleration have directions associated with them. By adding or combining such quantities (**vector addition**) we can obtain the net effect that occurs—the *resultant*. Either the geometric method or analytic component method can be used to add two or more vectors.

22

© 2007 Pearson Education, Inc., Upper Saddle River, NJ. All rights reserved. This material is protected under all copyright laws as they currently exist. No portion of this material may be reproduced, in any form or by any means, without permission in writing from the publisher.

Vector Addition: Geometric Methods
Triangle Method

To add two vectors $\vec{A}$ and $\vec{B}$ by the **triangle method**
(also called *tip-to-tail method*), place the tail of $\vec{B}$ at the
head of $\vec{A}$. The resultant vector (the vector sum) $\vec{R}$ is the
vector joining the tail of $\vec{A}$ to the head of $\vec{B}$, or
$\vec{R} = \vec{A} + \vec{B}$.

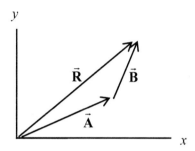

Measure the length of the resultant vector **R** (with a ruler)
for the magnitude and the angle of the vector with the
positive x axis (using a protractor) for the direction.
Change the length of the arrow to the magnitude of the
vector by multiplying with the chosen scale factor. This
method can be applied to more than two vectors, in which
case it is called the polygon method.

Vector Subtraction

An arrow having the same length as the original vector,
but pointing in the opposite direction, represents the
negative of a vector.

23

2007 Pearson Education, Inc., Upper Saddle River, NJ. All rights reserved. This material is protected under all copyright laws as they currently exist.
No portion of this material may be reproduced, in any form or by any means, without permission in writing from the publisher.

To subtract two vectors $\vec{A}$ and $\vec{B}$, add $\vec{A}$ with $-\vec{B}$. Thus,

$\vec{A} - \vec{B} = \vec{A} + (-\vec{B})$.

Vector Components and the Analytical Component Method

The graphical method of adding vectors is usually an approximation. Vectors can be added more accurately by the analytical method of vector components.

Adding Rectangular Vector Components

By rectangular components we mean those at right angles to each other. Suppose that the two vectors $\vec{A}$ and $\vec{B}$ are at right angles. Then the magnitude of the resultant vector $\vec{C}$ is given by the Pythagorean theorem:

$$C = \sqrt{A^2 + B^2} \qquad (3.4a)$$

The orientation of $\mathbf{C}$ with respect to the x axis is given by the inverse tangent relationship

$$\theta = \tan^{-1}\frac{B}{A} \qquad (3.4b)$$

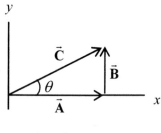

$$\vec{C} = \vec{A} + \vec{B}$$

24

© 2007 Pearson Education, Inc., Upper Saddle River, NJ. All rights reserved. This material is protected under all copyright laws as they currently exist. No portion of this material may be reproduced, in any form or by any means, without permission in writing from the publisher.

Resolving a Vector into Rectangular Components: Unit Vectors

Resolving a vector into rectangular components is the reverse of adding the components. If a vector $\vec{C}$ makes an angle θ relative to the positive x axis, the x and y components of the vector are given by, respectively,

$$C_x = C \cos \theta \qquad\qquad (3.5a)$$
$$C_y = C \sin \theta \qquad\qquad (3.5b)$$

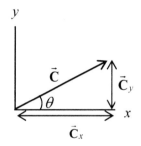

$$\vec{C} = \vec{C}_x + \vec{C}_y$$

C_x and C_y can be positive, negative, or zero. The direction of the vector can be obtained from the magnitude of the components and is given by

$$\theta = \tan^{-1} \frac{C_y}{C_x} \qquad\qquad (3.6)$$

An arbitrary vector can be conveniently expressed in terms of its components using **unit vectors**. The unit vector $\hat{x}$ is a dimensionless vector of magnitude unity pointing in the positive x direction. The unit vector $\hat{y}$ is a dimensionless vector of magnitude unity pointing in the

25

©2007 Pearson Education, Inc., Upper Saddle River, NJ. All rights reserved. This material is protected under all copyright laws as they currently exist. No portion of this material may be reproduced, in any form or by any means, without permission in writing from the publisher.

positive y direction. An arbitrary vector **C** can be written as the sum of its vector components as follows:

$$\vec{C} = C_x \hat{x} + C_y \hat{y} \tag{3.7}$$

Vector Addition Using Components

The **analytical component method** of vector addition involves the following procedure:

1. Resolve the vectors into x and y components. Use the acute angles between the vectors and the x-axis and indicate the directions of the components by using plus or minus signs.
2. Add all the x-components together and all the y-components together to get the x- and y-components of the resultant.
3. Express the resultant vector, using:

    a. the component form: $\vec{C} = C_x \hat{x} + C_y \hat{y}$

    b. the magnitude-angle form:
        The magnitude:
$$C = \sqrt{C_x^2 + C_y^2}, \text{ the angle}$$
$$\theta = \tan^{-1}\left|\frac{C_y}{C_x}\right|$$

The component can be extended to three dimensions.

© 2007 Pearson Education, Inc., Upper Saddle River, NJ. All rights reserved. This material is protected under all copyright laws as they currently exist. No portion of this material may be reproduced, in any form or by any means, without permission in writing from the publisher.

3.3 Projectile Motion

A **projectile motion** is a motion under the action of gravity only. A special case of projectile motion occurs in one dimension when an object is projected vertically upward (free fall).

Horizontal Projections
We consider the x- axis to be horizontal and the y-axis to be vertical, with the positive direction upward. Acceleration in the y direction is $a_y = -g$. Since gravity causes no acceleration in the x direction, $a_x = 0$. For a projectile launched horizontally, the angle θ between the initial velocity and the horizontal is zero. In this case, $v_{x0} = v_0 \cos 0^0 = v_0$ and $v_{y0} = v_0 \sin 0^0 = 0$. Thus,

$v_x = v_{x0}$, which shows that the velocity in the x direction does not change.

$$x = v_{x0}t$$

For the y direction

$$v_y = -gt$$

Thus, the object travels at a uniform velocity in the horizontal direction while *at the same time* undergoing acceleration in the downward direction under the influence of gravity. The time of flight of the object is *exactly the same as if it were falling vertically.*

Projection at Arbitrary Angles
A projectile that is launched at an angle θ with respect to the horizontal has the following initial velocity components:

$$v_{x0} = v_0 \cos \theta \qquad (3.8a)$$
$$v_{y0} = v_0 \sin \theta \qquad (3.8b)$$

©2007 Pearson Education, Inc., Upper Saddle River, NJ. All rights reserved. This material is protected under all copyright laws as they currently exist. No portion of this material may be reproduced, in any form or by any means, without permission in writing from the publisher.

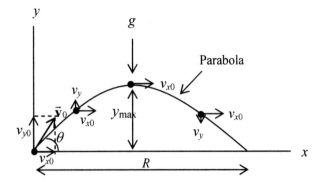

Since there is no horizontal acceleration and gravity acts in the negative y direction:

$$v_x = v_{x0} = v_0 \cos \theta \qquad (3.9a)$$
$$v_y = v_0 \sin \theta - gt \qquad (3.9b)$$

The displacements in the x and y directions are given, respectively:

$$x = (v_0 \cos \theta) t \qquad (3.10a)$$

$$y = (v_0 \sin \theta)t - \frac{1}{2} gt^2 \qquad (3.10b)$$

The curve produced by the preceding equations, or the path of a projectile, is a parabola.

The maximum horizontal distance that a projectile travels before it lands is called the *range*, R, of the projectile. To determine range, we first show that the time to reach the maximum height is

$$t_{max} = \frac{v_0 \sin \theta}{g}$$

Then,

© 2007 Pearson Education, Inc., Upper Saddle River, NJ. All rights reserved. This material is protected under all copyright laws as they currently exist. No portion of this material may be reproduced, in any form or by any means, without permission in writing from the publisher.

$$R = v_x t = (v_0 \cos \theta)(2t_{max}) = \frac{v_0^2 \sin 2\theta}{g} \qquad (3.11)$$

Range R depends on the initial speed and the angle of projection. For the maximum value of R, $\sin 2\theta = 1$, and $\theta = 45°$. Thus,

$$R_{max} = \frac{v_0^2}{g} \qquad (3.12)$$

Thus, for a maximum range, a projectile should *ideally* be projected at 45°. However, when air resistance is a factor, the angle of projection for maximum range is less than 45°. Other factors such as spin and wind may also affect the range.

Projectile motion exhibits striking symmetries. Because of symmetry, the range is the same for angles equally above and below 45°, such as for 30° and 60°. In the path of a projectile, the time a projectile takes to reach its highest point is half of the total time of flight. At any given height, the speed of the projectile is the same on the way up as on the way down. Symmetry considerations are useful for solving problems.

3.4 Relative Velocity

When an object is in motion in a reference frame that is also in motion, the motion of the object will generally appear different to an observer at rest outside the reference frame versus an observer at rest in the reference frame. For example, the motion of an object inside a moving train will appear different to an observer at rest on the ground than to a passenger at rest in the train. Hence, we say that

29

2007 Pearson Education, Inc., Upper Saddle River, NJ. All rights reserved. This material is protected under all copyright laws as they currently exist. No portion of this material may be reproduced, in any form or by any means, without permission in writing from the publisher.

motion is *relative* to some reference frame, introducing the concept of **relative velocity**.

Relative Velocities in One Dimension

Relative velocities can be found by vector subtraction, when the two velocities are linear in the same or opposite directions and all have the same reference. For example, the velocity of a car B relative to car A is given by

$$\vec{\mathbf{v}}_{BA} = \vec{\mathbf{v}}_B - \vec{\mathbf{v}}_A$$

where $\vec{\mathbf{v}}_A$ and $\vec{\mathbf{v}}_B$ are velocities of the cars A and B with respect to the ground, respectively.

For a passenger walking on a moving walkway, the velocity of a passenger relative to the ground, $\vec{\mathbf{v}}_{pg}$, is given by

$$\vec{\mathbf{v}}_{pg} = \vec{\mathbf{v}}_{pw} + \vec{\mathbf{v}}_{wg}$$

where

$\vec{\mathbf{v}}_{pw}$ = velocity of the passenger relative to the walkway, and $\vec{\mathbf{v}}_{tg}$ = velocity of the walkway relative to the ground. If the walkway moves in the opposite direction at the same speed as the passenger moves, $\vec{\mathbf{v}}_{pw} = -\vec{\mathbf{v}}_{wg}$. $\vec{\mathbf{v}}_{pg} = 0$.

Relative Velocities in Two Dimensions

For two-dimensional motions, use the rectangular components to add or subtract vectors to determine relative velocities.

© 2007 Pearson Education, Inc., Upper Saddle River, NJ. All rights reserved. This material is protected under all copyright laws as they currently exist. No portion of this material may be reproduced, in any form or by any means, without permission in writing from the publisher.

Hints and Suggestions for Solving Problems

1. Two-dimensional motion is equivalent to *two* one-dimensional motions independent of each other. Motion in each direction can be studied using the equations described in Chapter 2.

2. To find the components of a vector $\vec{C}$ use $C_x = C \cos \theta$ and $C_y = C \sin \theta$. Remember that θ is the angle made by the vector with the positive x-axis.

3. Given the components of a vector, find the vector by using the Pythagorean theorem (for its magnitude) and the inverse tangent of the ratio of its y and x components (for its direction). Problems in Section 3.1 can be solved using tips 1 and 2.

4. To add (or subtract) two vectors by components, simply add (or subtract) the x components of the individual vectors to find the x component of the resultant vector. Add the y components of the individual vectors to find the y component of the resultant vector. Then use the Pythagorean theorem and the inverse tangent of the ratio of the y and x components of the resultant to find the magnitude and direction of the resultant, respectively. Use this technique to solve problems in Section 3.2.

5. Make sure you know how to express a vector using unit vector notations, $\vec{C} = C_x \hat{\mathbf{x}} + C_y \hat{\mathbf{y}}$, and how to add or subtract vectors using unit vector notations.

31

©2007 Pearson Education, Inc., Upper Saddle River, NJ. All rights reserved. This material is protected under all copyright laws as they currently exist. No portion of this material may be reproduced, in any form or by any means, without permission in writing from the publisher.

6. Consider the right and upward directions as positive directions. Any vector (such as velocity and acceleration) directed right or upward is considered positive. Any vector directed left or downward will have a negative sign before its magnitude.

7. For projectile motion there is acceleration $a = -g = -9.80 \text{ m/s}^2$ in the y direction. There is no acceleration in the x direction. Thus, $v_{x0} = v_x = v_0$.

8. For a projectile launched horizontally, $v_{x0} = v_0$ and $v_{y0} = 0$. Make these substitutions to solve problems such as those in Section 3.3.

9. For a projectile launched with an arbitrary angle θ, its initial x and y components of velocity are $v_{x0} = v_0 \cos \theta$ and $v_{y0} = v_0 \sin \theta$, respectively. Make these substitutions to solve problems such as those in Section 3.3.

10. Take advantage of symmetry when solving projectile motion problems. Because of symmetry in projectile motion, the time it takes to reach the maximum height is one half of the total time of flight.

11. At the maximum height, velocity is in the x direction, that is, $v_y = 0$. Use this fact to determine the *maximum height*.

12. When a projectile reaches the ground, $y = 0$. Use this fact to determine its *time of flight*. Find the *range* by multiplying the time of flight by the speed in the x direction. It is *not* necessary to remember the expressions for the range or the time of flight. You should be able to derive the expressions when necessary.

32

© 2007 Pearson Education, Inc., Upper Saddle River, NJ. All rights reserved. This material is protected under all copyright laws as they currently exist. No portion of this material may be reproduced, in any form or by any means, without permission in writing from the publisher.

CHAPTER 4
FORCE AND MOTION

This chapter describes Newton's three laws of motion, which describe the dynamics of linear motion. The concepts of force, weight, normal force, translational equilibrium, and friction are introduced.

Key Terms and Concepts

> Force
> Net force
> Contact and action-at-distance forces
> Inertia
> Newton's first, second, and third laws of motion
> Newton
> Weight
> Free-body diagram
> Translational equilibrium
> Friction
> Static, kinetic, and rolling frictions
> Coefficients of static and kinetic frictions
> Air resistance
> Terminal velocity

4.1 The Concepts of Force and Net Force

Force is simply a push or a pull. Two quantities that characterize a force are its *magnitude* and the *direction* in which it is applied. Force is a vector quantity. Force can produce a change in motion or velocity of an object. The total force or **net force,** $\vec{F}_{net}$, exerted on an object is the vector sum of the individual forces ($\Sigma \vec{F}_i$) acting on it.

33

©2007 Pearson Education, Inc., Upper Saddle River, NJ. All rights reserved. This material is protected under all copyright laws as they currently exist. No portion of this material may be reproduced, in any form or by any means, without permission in writing from the publisher.

Forces are divided into two types. *Contact forces* are those that require physical contact between objects, such as the force you apply when you push a door or kick a ball. *Action-at-a-distance* forces are those that do not require any physical contact, such as the force of gravity between two masses or the electric force between two charges.

4.2 Inertia and Newton's First Law of Motion

Newton's first law is a modified statement of the *law of inertia*, given originally by Galileo. Galileo observed from his experiments that an object tends to maintain its state of motion. The natural tendency of an object to maintain its state of rest or its uniform velocity is called its **inertia**. Newton related the concept of inertia to mass. The larger the mass of an object, the more it resists changes of motion. Mass is, thus, a measurement of the inertia of an object.

Newton's first law of motion states that in the absence of any unbalanced applied force (that is, $\vec{\mathbf{F}}_{net} = 0$), a body at rest remains at rest and a body in motion continues its motion with the same velocity.

4.3 Newton's Second Law of Motion

Newton's second law provides a relationship between the cause (force) and the effect (acceleration) of a motion. It states that when a net force $\vec{\mathbf{F}}_{net}$ acts on an object of mass m, the acceleration $\vec{\mathbf{a}}$ of the object will be given by

$$\vec{\mathbf{F}}_{net} = m\vec{\mathbf{a}} \qquad (4.1)$$

34

© 2007 Pearson Education, Inc., Upper Saddle River, NJ. All rights reserved. This material is protected under all copyright laws as they currently exist. No portion of this material may be reproduced, in any form or by any means, without permission in writing from the publisher.

The SI unit of force is the newton (N),
$$1 \text{ N} = 1 \text{ kg} \cdot \text{m/s}^2$$

Weight

The **weight** of an object is the amount of gravitational force exerted by the Earth on the object. For example, if an object is dropped, the net force acting on it is its weight, w, and its acceleration is the acceleration due to gravity, g. Thus, from Newton's second law (Equation 4.1),

$$w = mg \qquad (4.2)$$

It is clear from this equation that the weight of 1.0 kg mass on the Earth is 9.8 N (since $g = 9.8$ m/s^2). Keep in mind that *the mass is a fundamental property*, but weight depends on the value of g, and, as a result it is different where g is different (such as on the moon where g is about 1/6 that on Earth).

The Second Law in Component Form

For a two-dimensional motion, Newton's second law applies to each component of motion. In terms of components

$$\Sigma F_x = ma_x \text{ and } \Sigma F_y = ma_y \qquad (4.3)$$

where ΣF_x and ΣF_y are net forces in x and y directions and a_x and a_y are corresponding components of the acceleration, respectively.

4.4 Newton's Third Law of Motion

Forces always occur in pairs. For every force (action) that acts on an object there is an equal and opposite force

2007 Pearson Education, Inc., Upper Saddle River, NJ. All rights reserved. This material is protected under all copyright laws as they currently exist. No portion of this material may be reproduced, in any form or by any means, without permission in writing from the publisher.

(reaction). This is **Newton's third law**. In symbolic notation,

$$\vec{F}_{12} = -\vec{F}_{21}$$

where $\vec{F}_{12}$ is the force exerted *on* object 1 *by* object 2 and $\vec{F}_{21}$ is the force exerted on object 2 by object 1.

Action–reaction forces *do not act on the same object*; hence, they do *not* cancel. Action–reaction forces generally produce different accelerations, since the masses of the objects are likely to be different.

For the motion of a rocket, rocket and exhaust gases exert equal and opposite forces on each other. The exhaust gases are accelerated away from the rocket, whereas the rocket is accelerated in the opposite direction.

When an object rests on a surface, the surface provides a force on the object in a direction *perpendicular* to the surface. This is called the **normal force**, $\vec{N}$. For an object on a horizontal surface, the normal force is vertical. In this case, from Newton's second law,

$$\Sigma F_y = N - mg = ma_y = 0,$$

or $N = mg$

© 2007 Pearson Education, Inc., Upper Saddle River, NJ. All rights reserved. This material is protected under all copyright laws as they currently exist. No portion of this material may be reproduced, in any form or by any means, without permission in writing from the publisher.

Normal force is not always equal and opposite to the weight of an object. For an inclined plane, the normal force is still at right angles to the surface, but not in the vertical direction. For an object of weight mg resting on an inclined plane of angle θ,

$$\Sigma F_y = N - mg \cos \theta = ma_y = 0,$$

or $N = mg \cos \theta$.

4.5 More on Newton's Laws: Free-Body Diagrams and Translational Equilibrium

When solving problems involving force and Newton's second law, it is convenient to draw a free-body diagram, showing *all external forces* acting on the object. In a free-body diagram, the vectors do not have to be drawn to scale, but whether there is a net force in any direction or whether the forces are balanced should be apparent from the diagram.

The usual steps in constructing a free-body diagram are as follows:

37

2007 Pearson Education, Inc., Upper Saddle River, NJ. All rights reserved. This material is protected under all copyright laws as they currently exist. No portion of this material may be reproduced, in any form or by any means, without permission in writing from the publisher.

1. Make a sketch and identify the forces.
2. Isolate the object of interest and draw the Cartesian axes.
3. Draw properly oriented force vectors from the origin.
4. Resolve the forces that are not directed along the axes.

Problem-Solving Hint
- Draw a free-body diagram for each object in the system.
- Resolve the forces into x and y components, if necessary.
- Apply Newton's second law either to the system as a whole or to a part of the system, depending on the situation.

Translational Equilibrium
When the net force acting on an object is zero, the object is said to be in **translational equilibrium**,

$$\Sigma \vec{F}_i = 0 \qquad (4.4)$$

This means the object either remains at rest or moves with a constant velocity.

In component forms,

$$\Sigma \vec{F}_x = 0 \qquad (4.5)$$
$$\Sigma \vec{F}_y = 0$$

The preceding equations give what is called the **condition for translational equilibrium**.

© 2007 Pearson Education, Inc., Upper Saddle River, NJ. All rights reserved. This material is protected under all copyright laws as they currently exist. No portion of this material may be reproduced, in any form or by any means, without permission in writing from the publisher.

4.6 Friction

Friction is a resistance that opposes the relative motion between two materials or media in contact with each other. This resistance occurs in solids, liquids, and gases and is characterized by the *force of friction* (*f*). It was originally thought that friction was primarily due to the mechanical interlocking of surface irregularities but research has shown that the friction between two surfaces of contact in solids (particularly in metals) is primarily due to local adhesion.

In some situations friction should be reduced or eliminated, whereas in some other situations friction can be helpful.

There are three types of friction—static, sliding (kinetic), and rolling.

Static friction tends to keep the two surfaces in contact from moving relative to each other. **Sliding friction** or **kinetic friction** is the friction encountered when surfaces slide against one another with a finite speed. **Rolling friction** occurs when one surface rotates as it moves over another surface without slipping or sliding at the point or the area of contact.

Frictional Forces and Coefficients of Friction

It was found that the force of friction (static and kinetic) depends on the nature of the surfaces and the normal force, N. As the applied force increases, the force of static friction increases. The force of static friction, f_s, is always equal but oppositely directed to the applied force until the body is about to move, in which case the force of static friction attains its maximum value.

39

007 Pearson Education, Inc., Upper Saddle River, NJ. All rights reserved. This material is protected under all copyright laws as they currently exist. No portion of this material may be reproduced, in any form or by any means, without permission in writing from the publisher.

$$f_s \leq \mu_s N \qquad (4.6)$$

where the constant μ_s is called the **coefficient of static friction**.

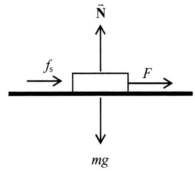

The maximum value of the force of static friction is:

$$f_{s,\,max} = \mu_s N \qquad (4.7)$$

Once an object is sliding, the force of kinetic friction acts in the opposite direction of motion and is given by

$$F_k = \mu_k N \qquad (4.8)$$

where the constant μ_k is called the **coefficient of kinetic friction**.

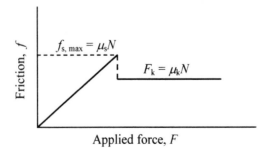

40

© 2007 Pearson Education, Inc., Upper Saddle River, NJ. All rights reserved. This material is protected under all copyright laws as they currently exist. No portion of this material may be reproduced, in any form or by any means, without permission in writing from the publisher.

Generally, the coefficient of kinetic friction is less than the coefficient of static friction.

Air Resistance

Air resistance is a frictional force that comes into play when an object moves through air. This arises because of the collision of the moving object with air molecules. Air resistance depends on the object's size and shape and its speed. From Newton's second law, $mg - f = ma$. For a falling object, the retarding force increases as the object accelerates. When the retarding force becomes the same as the weight of the object, the net force on the object is zero, and the object falls with a maximum constant velocity, called the **terminal velocity, v_t.**

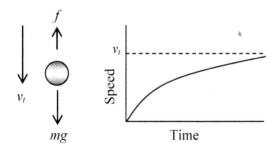

For a skydiver with an unopened parachute, the terminal velocity is about 125 mi/h or 200 km/h. When the parachute is open, giving a larger exposed area and a shape that catches the air, additional air resistance slows the diver down to reach a terminal velocity of about 25 mi/h (40 km/h), which is suitable for landing.

41

2007 Pearson Education, Inc., Upper Saddle River, NJ. All rights reserved. This material is protected under all copyright laws as they currently exist. No portion of this material may be reproduced, in any form or by any means, without permission in writing from the publisher.

Hints and Suggestions for Solving Problems

1. Begin solving a problem with a sketch. Isolate each object in the problem and draw a free-body diagram for each object of interest, showing all external forces with their directions.

2. Establish a convenient coordinate system. Find $\Sigma \vec{\mathbf{F}}$ for each component. Apply Newton's second law of motion separately for each component. (Equation 4.3b expresses Newton's second law in terms of components.) Keep this in mind when solving problems such as those in Section 4.3.

3. When adding several forces or several components of forces, make sure you add them with their proper signs. Any force directed toward the right or directed upward is considered to be positive. Any force directed toward the left or directed downward is considered to be negative.

4. Normal force is always normal to the surface. The force of gravity always acts vertically downward.

5. There is no acceleration in the normal direction. The sum of the normal components of the forces must be zero.

6. For an object moving on a flat surface the normal force and the weight are equal and opposite. For an object on an inclined plane, the magnitudes are different.

© 2007 Pearson Education, Inc., Upper Saddle River, NJ. All rights reserved. This material is protected under all copyright laws as they currently exist. No portion of this material may be reproduced, in any form or by any means, without permission in writing from the publisher.

7. If you need to resolve forces, keep in mind that the x and y directions may not always be the simplest directions along which to resolve the forces. For inclined surfaces, it is convenient to choose x and y axes parallel and perpendicular to the inclined surface, respectively.

8. A pulley changes the direction of a tension force without changing its magnitude.

9. Action–reaction forces always act on two *different* bodies, and they *never cancel* each other.

10. For an object in translational equilibrium, $\Sigma F_x = 0$ and $\Sigma F_y = 0$. Remember these equations when applying Newton's second law to solve problems on translational equilibrium such as those in Section 4.5.

11. Remember that friction force always opposes the motion and μ_s is greater than μ_k.

2007 Pearson Education, Inc., Upper Saddle River, NJ. All rights reserved. This material is protected under all copyright laws as they currently exist. No portion of this material may be reproduced, in any form or by any means, without permission in writing from the publisher.

CHAPTER 5
WORK AND ENERGY

This chapter provides a definition of work, discusses the work–energy theorem, defines kinetic and potential energies, and states a fundamental principle of nature called the conservation of energy. Conservative and nonconservative forces as well as power are also defined.

Key Terms and Concepts

Work done by a constant force
Positive and negative work
Joule
Work done by a variable force
Hooke's law
Spring constant
Work–energy theorem
Kinetic energy and potential energy
Conservation of energy
Conservative and nonconservative forces
Mechanical energy and its conservation
Power
Watt and Horsepower
Efficiency

5.1 Work Done by a Constant Force

Work, in everyday language, is said to be done when a task is accomplished. However, in physics the word *work* is related to the motion of an object and needs to be precisely defined.

When we push a shopping cart or pull a suitcase, we do work. In general, the greater the force, the greater the

© 2007 Pearson Education, Inc., Upper Saddle River, NJ. All rights reserved. This material is protected under all copyright laws as they currently exist. No portion of this material may be reproduced, in any form or by any means, without permission in writing from the publisher.

work, and also the greater the displacement, the greater the work. For a constant force, $\vec{F}$, acting in the same direction as the displacement, $\vec{d}$, of an object, the **work**, W, is given by,

$$W = Fd \qquad (5.1)$$

Obviously, W is zero if the displacement d is zero.

In some situations the direction of the applied force and the direction of motion of the object are different. For example, when a person pulls a suitcase on a level surface with a strap that makes an angle θ with the horizontal, the force is at an angle θ to the direction of motion.

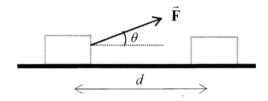

In this case, only the component of the force that is in the direction of motion of the suitcase does the work. Thus, when the angle between $\vec{F}$ and $\vec{d}$ is θ, is

$$W = F_{\parallel}d = (F \cos \theta)d \qquad (5.2)$$

The perpendicular component of the force ($F \sin \theta$), does no work since there is no displacement in this direction.

The SI unit of work is newton·meter, N·m, or joule (J). The British standard unit of work is *foot-pound* (ft·lb).

Work is a *scalar* quantity and can be positive (> 0) or negative (< 0).

007 Pearson Education, Inc., Upper Saddle River, NJ. All rights reserved. This material is protected under all copyright laws as they currently exist. No portion of this material may be reproduced, in any form or by any means, without permission in writing from the publisher.

$W > 0$ when $-90° < \theta < 90°$,
$W < 0$ when $90° < \theta < 270°$,
$W = 0$ when $\theta = \pm 90°$.

It is common to specify which force is doing work *on* which object. When you lift an object, *you* do work *on* the object. This is also described as doing work *against* gravity, since gravity acts in the direction opposite to the applied force. When more than one force acts on an object, the *total* or *net work* is the scalar sum of the work done by individual forces.

5.2 Work Done by a Variable Force

Most forces in nature vary with position. Examples include the force exerted by a spring and the force of gravity between a planet and the sun.

For small deformations, the relationship between the spring force F_s and the amount of stretch (or compression) is given by

$$F_s = -kx \qquad (5.3)$$

where k is a constant of the spring called the **spring constant** or **force constant**. The greater the value of k, the stiffer is the spring. This relation is known as **Hooke's law**. For large deformations Hooke's law is not valid.

In general, calculation of work done by variable forces requires calculus. However, the work done in stretching (or compressing) a spring can be calculated by using the average force, which is given by

46

© 2007 Pearson Education, Inc., Upper Saddle River, NJ. All rights reserved. This material is protected under all copyright laws as they currently exis
No portion of this material may be reproduced, in any form or by any means, without permission in writing from the publisher.

$$\overline{F} = \frac{F + F_0}{2} = \frac{F}{2} \text{ (since } F_0 = 0)$$

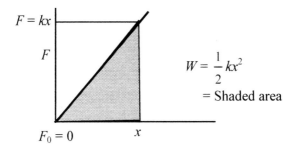

$$W = \frac{1}{2} kx^2$$
$$= \text{Shaded area}$$

Therefore,

$$W = \overline{F} x = \frac{1}{2} (kx) x = \frac{1}{2} kx^2 \tag{5.4}$$

Clearly, the work done is the shaded area under the curve.

5.3 The Work–Energy Theorem: Kinetic Energy

When positive work is done on an object, its speed increases, and when negative work is done on an object, its speed decreases. The work–energy theorem relates the work done with the change of speed of an object.

The **kinetic energy** of an object is the energy of its motion. For an object of mass m moving with a speed v, its kinetic energy K is defined as

$$K = \frac{1}{2} mv^2 \tag{5.5}$$

The SI unit of energy is the joule (J).

47

2007 Pearson Education, Inc., Upper Saddle River, NJ. All rights reserved. This material is protected under all copyright laws as they currently exist. No portion of this material may be reproduced, in any form or by any means, without permission in writing from the publisher.

Kinetic energy is proportional to the square of the speed of a moving object and thus is always positive.

The **work–energy theorem** states that the total work done on an object is equal to the change in its kinetic energy.

$$W = K - K_0 = \Delta K = \frac{1}{2} mv^2 - \frac{1}{2} mv_0^2 \qquad (5.6)$$

The work–energy theorem holds for any force and is one of the fundamental results in physics.

5.4 Potential Energy

Potential energy (U) is the energy stored in an object by virtue of its position and/or configuration. When a ball is lifted from the ground to a new height, the work done is stored in the form of potential energy. When the ball is released, gravity does work, and the potential energy of the ball decreases as it falls down.

Since the work done in stretching (or compressing) a spring from its equilibrium position is $W = \frac{1}{2} kx^2$, the potential energy stored in the spring is

$$U = W = \frac{1}{2} kx^2 \qquad (5.7)$$

if we choose the potential energy at the equilibrium length of the spring, U_0 is zero.

Gravitational potential energy refers to the energy stored in an object because of its height above some reference point, such as the floor or the ground.

48

© 2007 Pearson Education, Inc., Upper Saddle River, NJ. All rights reserved. This material is protected under all copyright laws as they currently exist. No portion of this material may be reproduced, in any form or by any means, without permission in writing from the publisher.

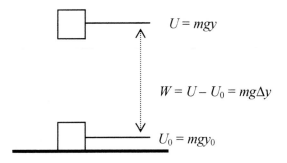

When an object of mass m is lifted by Δy, work is done, W, against the force of gravity ($F = mg$), and the work is equal to the change in potential energy, ΔU, of the object.

Work = change in potential energy

or $\qquad W = F\Delta y = \Delta U$,

or $\qquad mgh - mgh_0 = U - U_0$.

Thus, with the common choice of $h_0 = 0$ and $U_0 = 0$, the gravitational potential energy is

$$U = mgy \qquad (5.8)$$

Zero Reference Point

Potential energy is the energy of *position*, and the choice of reference position or point of zero potential energy is arbitrary, such as the choice of the origin of a coordinate system. A convenient choice for zero gravitational potential energy ($U_0 = 0$) is for $h_0 = 0$. However, the *difference* or *change* in potential energy associated with two positions is the same regardless of the choice of reference position or point.

The potential energy of an object can be negative since the choice of reference point is arbitrary. When an object has a negative potential energy, it is said to be in a potential

49

2007 Pearson Education, Inc., Upper Saddle River, NJ. All rights reserved. This material is protected under all copyright laws as they currently exist. No portion of this material may be reproduced, in any form or by any means, without permission in writing from the publisher.

energy *well*. Also note that when an object is moved from one height to another height, the change in gravitational potential energy is *independent* of the path of transition.

5.5 The Conservation of Energy

Conservation laws are fundamental principles of nature. The **law of conservation of energy** states that *the total energy of an isolated system is always conserved.* This means that energy cannot be created nor destroyed; it can only be transformed from one form to another.

Conservative and Nonconservative Forces
Forces are classified into two types: conservative and nonconservative.

Following are the definitions of a conservative force:

A force is said to be conservative if the work done by or against it in moving an object does not depend on the path of the object.

A force is conservative if the work done by or against it is zero in moving an object in a closed path.

The force of gravity and spring force are examples of conservative force. When a conservative force does work, the work is stored in the object in the form of energy that can be released at a later time. Lifting a box against gravity requires work, which is stored in the box in the form of gravitational potential energy. The work done depends on the change of height but not on the path. When the box is released, it falls down, and gravity does the same amount of work. Thus the total work done in a closed path for gravity is zero.

© 2007 Pearson Education, Inc., Upper Saddle River, NJ. All rights reserved. This material is protected under all copyright laws as they currently exist. No portion of this material may be reproduced, in any form or by any means, without permission in writing from the publisher.

Friction is a nonconservative force. Work done against friction *does* depend on path. The energy associated with the work done against friction is converted to heat. The work done against friction is not stored in the form of potential energy.

Conservation of Total Mechanical Energy
The **total mechanical energy**, E, of an object is defined as the sum of its potential and kinetic energies.

$$E = K + U \qquad (5.9)$$

The **law of the conservation of mechanical energy** states: *If all the forces are conservative, the mechanical energy of an object is conserved throughout its motion.*
Thus,

$$\frac{1}{2}mv^2 + U = \frac{1}{2}mv_0^2 + U_0 \qquad (5.10)$$

This means that the kinetic and potential energies in a conservative system (where forces are all conservative) may change, but their sum is always constant.

For a conservative system, when work is done and energy is transferred within a system

$$\Delta K + \Delta U = 0 \qquad (5.11)$$

Total Energy and Nonconservative Forces
In reality, both conservative and nonconservative forces do work. In general, when work is done by conservative forces (W_c) and nonconservative forces (W_{nc})

$$W_c + W_{nc} = K - K_0 \qquad (5.12)$$

51

2007 Pearson Education, Inc., Upper Saddle River, NJ. All rights reserved. This material is protected under all copyright laws as they currently exist.
No portion of this material may be reproduced, in any form or by any means, without permission in writing from the publisher.

When nonconservative forces work, the mechanical energy of the system is not conserved. It can be shown from the preceding equation that the nonconservative work is equal to the change in mechanical energy.

$$W_{nc} = E - E_0 = \Delta E \qquad (5.13)$$

Note that in a nonconservative system the *total* energy is conserved, but not all of it is available for mechanical work.

5.6 Power

The time rate at which work is done is called **power**. If work W is done in time t, the average power is given by

$$\overline{P} = \frac{W}{t} \qquad (5.14)$$

For an object moving with a constant speed v,

$$\overline{P} = \frac{W}{t} = \frac{Fd}{t} = F\overline{v} \qquad (5.15)$$

If the force and the displacement are not in the same direction,

$$\overline{P} = \frac{Fd \cos\theta}{t} = F\overline{v} \cos\theta \qquad (5.16)$$

The SI unit of power is J/s also called a **watt** (W). A common unit of electric power is *kilowatt* (KW). The British unit of power is foot-pounds per second (ft·lb/s). Another common unit of power is **horsepower** (**hp**).
1 hp = 550 ft·lb/s = 746 W.

© 2007 Pearson Education, Inc., Upper Saddle River, NJ. All rights reserved. This material is protected under all copyright laws as they currently exist. No portion of this material may be reproduced, in any form or by any means, without permission in writing from the publisher.

Efficiency

Efficiency of a machine simply tells us the amount of *useful* work output for a given energy input. It is expressed by the fraction

$$\varepsilon = \frac{\text{work output}}{\text{energy input}} \times 100\% = \frac{W_{out}}{E_{in}} \times 100\% \qquad (5.17)$$

In terms of power,

$$\varepsilon = \frac{P_{out}}{P_{in}} \times 100\% \qquad (5.18)$$

Hints and Suggestions for Solving Problems

1. Work done by a constant force is $W = Fd \cos \theta$, where θ is the angle between the force and the displacement. The result can be positive, negative, or zero depending on θ.

2. When calculating work, make sure that the value of F is in newtons and d is in meters to result in W in joules. If $\theta = 0$, $W = Fd$. The expression for W will be useful for solving problems in Section 5.1.

3. For a variable force, the area between the force curve and the position axis (x-axis) gives the value of work. The result is consistent with the definition of work. This concept can be used to calculate the work done to stretch (or compress) a spring or work done by other variable forces, such as problems in Section 5.2.

4. Many problems in this chapter, such as those in Section 5.3, can be solved using the work–energy

53

2007 Pearson Education, Inc., Upper Saddle River, NJ. All rights reserved. This material is protected under all copyright laws as they currently exist. No portion of this material may be reproduced, in any form or by any means, without permission in writing from the publisher.

theorem: $W = \Delta K = \frac{1}{2} mv^2 - \frac{1}{2} mv_0^2$. Use the theorem to calculate change in kinetic energy or change in speed as a result of work done. When using this theorem, make sure that W is the *total work* with the *proper sign*.

5. When computing change in kinetic energy, remember that $v^2 - v_0^2$ is *not* equal to $(v - v_0)^2$.

6. To calculate gravitational potential energy, such as those in Section 5.4, choose a horizontal level for which $y_0 = 0$, then use $U = mgy$.

7. Remember that gravity is a conservative force and that the work done in a gravitational field is given by the change in gravitational potential energy.

8. The principle of conservation of mechanical energy can be used to solve many problems in this chapter, such as those in Section 5.5. According to this principle, the total mechanical energy $E = U + K$ is a constant if nonconservative forces are absent.

9. To calculate potential energy stored in a spring use $U = \frac{1}{2} kx^2$. You can apply energy conservation in a system involving the spring.

10. Work done by nonconservative forces is given by the relation $W_{nc} = E - E_0$ (the change in energy of the object).

11. To calculate power, such as problems in Section 5.6, use either $P = W/t$, or $F\bar{v}$, or $F\bar{v} \cos \theta$, depending on the problem. Use the conversion 1 hp = 746 W where necessary.

© 2007 Pearson Education, Inc., Upper Saddle River, NJ. All rights reserved. This material is protected under all copyright laws as they currently exist. No portion of this material may be reproduced, in any form or by any means, without permission in writing from the publisher.

CHAPTER 6
LINEAR MOMENTUM AND COLLISIONS

This chapter introduces linear momentum and impulse, and describes conservation of linear momentum. Two types of collisions are discussed. The concepts of center of mass, center of gravity, and motion of rockets are described.

Key Terms and Concepts

Linear momentum
Newton's second law
Impulse
Impulse–momentum theorem
Conservation of linear momentum
Elastic and inelastic collisions
Center of mass and center of gravity
Jet propulsion
Reverse thrust

6.1 Linear Momentum

The **linear momentum** (or simply momentum), $\vec{\mathbf{p}}$, is a characteristic of a moving object. It is defined as the product of the mass m and the velocity $\vec{\mathbf{v}}$ of the object:

$$\vec{\mathbf{p}} = m\,\vec{\mathbf{v}} \tag{6.1}$$

The SI unit of momentum is kilogram-meters per second or kg · m/s. Linear momentum is a vector quantity.

In terms of components,
$$p_x = mv_x$$

55

2007 Pearson Education, Inc., Upper Saddle River, NJ. All rights reserved. This material is protected under all copyright laws as they currently exist. No portion of this material may be reproduced, in any form or by any means, without permission in writing from the publisher.

$$p_y = mv_y$$

The **total linear momentum** of a system of objects is the vector sum of the momenta of the individual objects:

$$\vec{P} = \vec{p}_1 + \vec{p}_2 + \vec{p}_3 + \dots = \sum \vec{p}_i \qquad (6.2)$$

Force and Momentum

Change in momentum of an object requires a force. This is because momentum is directly related to the velocity and a change in velocity requires a force.

From Newton's second law, we can write

$$\vec{F}_{net} = m\vec{a} = \frac{m(\vec{v} - \vec{v}_0)}{\Delta t} = \frac{m\vec{v} - m\vec{v}_0}{\Delta t} = \frac{\vec{p} - \vec{p}_0}{\Delta t}$$

or, $$\vec{F}_{net} = \frac{\Delta \vec{p}}{\Delta t} \qquad (6.3)$$

Thus, *the time rate of change of momentum of an object is equal to the net external force on the object.* This is Newton's second law in terms of momentum.

6.2 Impulse

Real forces vary with time. For example, the force between a golf club and a golf ball rises rapidly to a large value, then falls back to zero. Physicists usually describe such a situation in terms of the average force, $\vec{F}_{avg}$.

56

© 2007 Pearson Education, Inc., Upper Saddle River, NJ. All rights reserved. This material is protected under all copyright laws as they currently exist. No portion of this material may be reproduced, in any form or by any means, without permission in writing from the publisher.

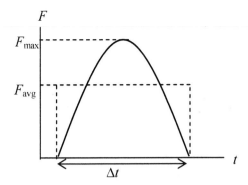

Thus, from Newton's second law:

$$\vec{F}_{avg} = \frac{\Delta \vec{p}}{\Delta t}$$

or $\quad \vec{F}_{avg} \Delta t = \Delta \vec{p} = \vec{p} - \vec{p}_0 \qquad (6.4)$

The quantity $\overline{F} \, \Delta t$ is called the **impulse** of the force.

$$\text{Impulse} = \vec{F}_{avg} \, \Delta t = \Delta \vec{p} = m\vec{v} - m\vec{v}_0 \quad (6.5)$$

Thus, the impulse exerted on a body is equal to its change of momentum. This is called **impulse–momentum theorem**.

The SI unit of impulse is N·s, which is also the unit of momentum. 1 kg · m/s = 1 N·s. Impulse is a vector that points in the same direction as the average force.

A collision is an interaction between objects in which there is an exchange of momentum and energy. Kinetic energy (K) can be expressed in terms of momentum (p) as

57

2007 Pearson Education, Inc., Upper Saddle River, NJ. All rights reserved. This material is protected under all copyright laws as they currently exist.
No portion of this material may be reproduced, in any form or by any means, without permission in writing from the publisher.

$$K = \frac{p^2}{2m} \qquad (6.6)$$

6.3 The Conservation of Linear Momentum

If the net force acting on an object is zero, then

$$\vec{\mathbf{F}}_{net} = \frac{\Delta \vec{\mathbf{p}}}{\Delta t} = 0,$$

or $\qquad \Delta \vec{\mathbf{p}} = 0 = \vec{\mathbf{p}} - \vec{\mathbf{p}}_0$

Thus, $\qquad \vec{\mathbf{p}} = \vec{\mathbf{p}}_0$

or $\qquad m\vec{\mathbf{v}} = m\vec{\mathbf{v}}_0 \qquad (6.7)$

Thus, the momentum of the object is conserved. The conservation of momentum also holds for a system of particles. A generalized statement of the **conservation of linear momentum** is: *If the net external force acting on a system is zero, the total linear momentum of the system is conserved.*

Conservation of momentum is one of the most fundamental principles in physics and is used to analyze collisions of objects from subatomic particles to astronomical objects.

6.4 Elastic and Inelastic Collisions

When two objects strike each other, a **collision** occurs, in which case the net external force is zero (or negligibly small). Collisions are classified into two types, *elastic* and *inelastic*, on the basis of what happens to the kinetic energy.

58

© 2007 Pearson Education, Inc., Upper Saddle River, NJ. All rights reserved. This material is protected under all copyright laws as they currently exist. No portion of this material may be reproduced, in any form or by any means, without permission in writing from the publisher.

If the total kinetic energy is conserved in a collision, the collision is called **elastic**.

Total K after = total K before

$$K_f = K_i \qquad (6.8)$$

If the total kinetic energy is *not* conserved in a collision, the collision is called inelastic. In this collision, work is done by conservative forces (such as friction) and some kinetic energy is lost.

Total K after < total K before

$$K_f < K_i \qquad (6.9)$$

For an isolated system, the total momentum is conserved for both the elastic and inelastic collisions.

Momentum and Energy in Inelastic Collisions

In the special case of an inelastic collision in which the objects stick together after collision, the collision is called **completely inelastic**. The maximum amount of kinetic energy is lost in a completely inelastic collision.

Let us consider a completely inelastic collision of two balls. The ball of mass m_1 is initially moving with the speed v_0 while the other ball of mass m_2 is at rest. If both the balls stick together and move with speed v after the collision, then from momentum conservation,

$$m_1 v_0 = (m_1 + m_2)\, v$$

or $$v = \left(\frac{m_1}{m_1 + m_2} \right) v_0 \qquad (6.10)$$

59

2007 Pearson Education, Inc., Upper Saddle River, NJ. All rights reserved. This material is protected under all copyright laws as they currently exist. No portion of this material may be reproduced, in any form or by any means, without permission in writing from the publisher.

It can be shown that in this case the ratio of total final kinetic energy (K_f) to the total initial kinetic energy (K_i) is:

$$\frac{K_f}{K_i} = \frac{m_1}{m_1 + m_2} \tag{6.11}$$

Momentum and Energy in Elastic Collisions

Let two objects of mass m_1 and m_2 moving along the same line collide. Initially the first object was moving with speed v_{10} while the second object was moving with speed v_{20}. After collision the speeds of the objects are v_1 and v_2, respectively.

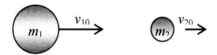

From the momentum conservation,

$$m_1 v_{10} + m_2 v_{20} = m_1 v_1 + m_2 v_2 \tag{6.12}$$

From the conservation of kinetic energy,

$$\frac{1}{2} m_1 v_{10}^2 + \frac{1}{2} m_2 v_{20}^2 = \frac{1}{2} m_1 v_1^2 + \frac{1}{2} m_2 v_2^2 \tag{6.13}$$

Using simple algebra we get:

$$v_{10} - v_{20} = -(v_1 - v_2)$$

© 2007 Pearson Education, Inc., Upper Saddle River, NJ. All rights reserved. This material is protected under all copyright laws as they currently exist.
No portion of this material may be reproduced, in any form or by any means, without permission in writing from the publisher.

$$v_1 = \left(\frac{m_1 - m_2}{m_1 + m_2} \right) v_{10} + \left(\frac{2 m_2}{m_1 + m_2} \right) v_{20} \quad (6.14)$$

$$v_2 = \left(\frac{2 m_1}{m_1 + m_2} \right) v_{10} - \left(\frac{m_1 - m_2}{m_1 + m_2} \right) v_{20} \quad (6.15)$$

One Object Initially at Rest

Let $v_{20} = 0$.

Case 1: If $m_1 = m_2$, $v_1 = 0$ and $v_2 = v_{10}$ (from Equation 6.15). This means that the objects have exchanged velocities as a result of the collision.

Case 2: If $m_1 \gg m_2$, $v_1 \approx v_{10}$ and $v_2 \approx 2v_{10}$. Thus, in the collision between a massive object and a stationary light object, the massive object is slowed down only slightly and the light object moves after collision with almost twice its initial speed.

Case 3: If $m_1 \ll m_2$, $v_1 \approx -v_{10}$ and $v_2 \approx 0$ (from Equation 6.15). Thus, in the collision between a stationary massive object and a light object, the massive object remains almost stationary and the light object recoils backward with almost the same speed.

6.5 Center of Mass

The **center of mass (CM)** is a point at which all of the mass of an object or system may be considered to be concentrated for the purpose of linear motion only. The center of mass is also described as the *balance point* of a mass. If a meterstick is balanced on your finger, its center

61

2007 Pearson Education, Inc., Upper Saddle River, NJ. All rights reserved. This material is protected under all copyright laws as they currently exist. No portion of this material may be reproduced, in any form or by any means, without permission in writing from the publisher.

of mass is located directly above the finger and all of the mass of the meterstick seems to concentrate at that point.

For a system of particles, if $\mathbf{a}_{CM}$ is the acceleration of its center of mass,

$$\vec{\mathbf{F}}_{net} = M\vec{\mathbf{A}}_{CM} \tag{6.16}$$

where M is the total mass of the system.

If the net external force on a system is zero, the total linear momentum of the system is conserved since

$$\vec{\mathbf{F}}_{net} = M\vec{\mathbf{A}}_{CM} = M\left(\frac{\Delta \vec{\mathbf{V}}_{CM}}{\Delta t}\right)$$

$$= \frac{\Delta\left(M\vec{\mathbf{V}}_{CM}\right)}{\Delta t} = \frac{\Delta\vec{\mathbf{P}}}{\Delta t} = 0 \tag{6.17}$$

This means that the center of mass either remains at rest or moves with a constant velocity, $\vec{\mathbf{V}}_{CM}$.

For a system in a plane containing objects of masses m_1, m_2, ..., at locations x_1, x_2, ..., the location of the center of mass is

$$X_{CM} = \frac{m_1 x_1 + m_2 x_2 + \cdots}{m_1 + m_2 + \cdots} \tag{6.18}$$

In shorthand notation, the x coordinate of the center of mass is

$$X_{CM} = \frac{\sum_i m_i x_i}{M} \tag{6.19}$$

© 2007 Pearson Education, Inc., Upper Saddle River, NJ. All rights reserved. This material is protected under all copyright laws as they currently exist. No portion of this material may be reproduced, in any form or by any means, without permission in writing from the publisher.

Center of Gravity

The center of gravity (CG) is a point where all the weight of an object may be considered to be concentrated in representing the object as a particle.

$$MgX_{CM} = \sum m_i gx_i \qquad (6.20)$$

6.6 Jet Propulsion and Rockets

Jet propulsion is the application of jets of gases, air, or liquid to the production of motion. It can be explained by Newton's third law and the conservation principle of linear momentum. A rocket is an example of jet propulsion. When the engine of a rocket is fired, it exhausts gases from burning fuel out the rear of the rocket. The escaping gases exert a net reaction force on the rocket that propels the rocket in the forward direction. The instantaneous momentum of the exhaust gases is equal and opposite to that of the rocket.

A rocket continuously loses mass as its fuel is burned. It is an example of a system of variable mass. As the mass of the rocket decreases, it accelerates more and more. A *reverse thrust* or braking thrust is needed to slow down the motion of a rocket (such as before landing on the moon). In this case an engine is fired to exhaust gases toward the motion, applying a reverse braking force on the rocket.

2007 Pearson Education, Inc., Upper Saddle River, NJ. All rights reserved. This material is protected under all copyright laws as they currently exist. No portion of this material may be reproduced, in any form or by any means, without permission in writing from the publisher.

Hints and Suggestions for Solving Problems

1. Draw a sketch of the problem. Remember that velocity and momentum are *vector* quantities. Take the direction of motion to the right as positive and stick with it for the whole problem.

2. The momentum vector is in the same direction as the velocity vector. The total momentum of a system of objects is the vector sum of the individual momenta of all the objects.

3. Determine the impulse from the change of momentum $\Delta \vec{\mathbf{p}}$ from its definition, $\vec{\mathbf{F}}_{avg} \Delta t = \Delta \vec{\mathbf{p}}$.

4. Determine whether the net external force is zero in the problem. If the net external force is zero, you can apply conservation of momentum ($\vec{\mathbf{p}} = \vec{\mathbf{p}}_0$) to solve problems, such as those in Section 6.3.

5. Apply the *momentum conservation principle* to solve collision problems such as those in Section 6.4. If the collision is elastic, apply also the fact that *total initial KE = total final KE*. It is not necessary to remember any equations for the final velocities in a collision.

6. Use Equation 6.19 to find the location of the center of mass of a system of particles.

7. Remember that the center of mass obeys Newton's second law of motion, $\vec{\mathbf{F}}_{net} = M \vec{\mathbf{A}}_{CM}$. This equation can be used to determine the motion of the center of mass.

64

© 2007 Pearson Education, Inc., Upper Saddle River, NJ. All rights reserved. This material is protected under all copyright laws as they currently exist. No portion of this material may be reproduced, in any form or by any means, without permission in writing from the publisher.

CHAPTER 7
CIRCULAR MOTION AND
GRAVITATION

This chapter defines the basic quantities for a circular motion and discusses centripetal acceleration and centripetal force. Newton's law of gravitation, Kepler's laws of planetary motion, the motion of satellites, and escape speed are discussed.

Key Terms and Concepts

Angular displacement
Radian
Angular speed and angular velocity
Period and frequency
Centripetal acceleration
Centrifuge
Centripetal force
Angular acceleration
Tangential acceleration
Newton's law of gravitation
Gravitational potential energy
Kepler's laws
Earth satellites
Escape speed
Weightlessness

7.1 Angular Measure

To describe circular motion it is convenient to define angular quantities that are analogous to position, velocity, and acceleration for linear motion.

2007 Pearson Education, Inc., Upper Saddle River, NJ. All rights reserved. This material is protected under all copyright laws as they currently exist.
No portion of this material may be reproduced, in any form or by any means, without permission in writing from the publisher.

For an object moving in a circle, its position can be designated by polar coordinates r and θ, which are related to the Cartesian coordinates by

$$x = r \cos \theta \qquad (7.1\text{a})$$
$$y = r \sin \theta \qquad (7.1\text{b})$$

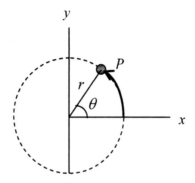

The **angular displacement** is given by an angle

$$\Delta\theta = \theta - \theta_0 \qquad (7.2)$$

or $\Delta\theta = \theta$ when we choose $\theta_0 = 0$.

A very common unit of angular displacement is the degree (°); there are 360° in one complete turn. An *arc length* is the distance traveled along the circle. The SI unit of θ is the radian (rad). A **radian** is defined as the angle for which the arc length (s) is equal to the radius (r) of the circle. For one complete revolution, $s = 2\pi r$; thus $\theta = 2\pi$ rad = 360°.

$$1 \text{ rad} = 57.3°$$

© 2007 Pearson Education, Inc., Upper Saddle River, NJ. All rights reserved. This material is protected under all copyright laws as they currently exist. No portion of this material may be reproduced, in any form or by any means, without permission in writing from the publisher.

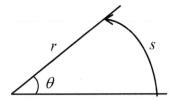

Clearly the arc length s, radius r, and angle θ in radians are related by

$$s = r\,\theta \qquad (7.3)$$

7.2 Angular Speed and Velocity

The angular speed is the magnitude of the time rate of change of angular displacement. The **average angular speed** ($\overline{\omega}$) is

$$\overline{\omega} = \frac{\Delta\theta}{\Delta t} = \frac{\theta - \theta_0}{t - t_0} \qquad (7.4)$$

Taking θ_0 and t_0 to be zero, we get

$$\theta = \overline{\omega}\,t = \omega\,t \quad \text{(if ω is constant)} \qquad (7.5)$$

The SI unit for angular speed is radian per second (rad/s or s^{-1}). A common unit of angular speed is revolutions per minute (rpm).

The **average and instantaneous angular velocities** are average and instantaneous speeds with directions. The direction of the angular velocity vector is given by the *right hand rule*, which states that when the fingers of your right hand are curled in the direction of circular motion, your extended thumb points toward the direction of ω. A

67

2007 Pearson Education, Inc., Upper Saddle River, NJ. All rights reserved. This material is protected under all copyright laws as they currently exist. No portion of this material may be reproduced, in any form or by any means, without permission in writing from the publisher.

circular motion is either clockwise or counterclockwise. The angular displacement is considered to be positive if the motion is counterclockwise.

Relationship between Tangential and Angular Speeds

For a particle moving in a circle, its instantaneous velocity is tangential to its circular path. If the motion is uniform, its **tangential speed** v is constant.

Since $s = r\,\theta = r(\omega\,\tau)$, and also, $s = vt$, we get

$$v = r\,\omega \tag{7.6}$$

This is the relation between the tangential speed and angular speed.

Note that particles rotating with constant angular velocity have the same angular speed, but their tangential speeds are different at different distances from the axis.

Period and Frequency

The time required to complete one revolution is called the **period**, T. The SI unit of period is second (s). The frequency, f, is the number of revolutions in one second. The SI unit of frequency is hertz (Hz, 1/s, or s^{-1}).

Clearly, $f = \dfrac{1}{T}$ $\qquad$ (7.7)

Since 2π radian is traveled in one period, we get

$$\omega = \frac{2\pi}{T} = 2\pi f \tag{7.8}$$

68

© 2007 Pearson Education, Inc., Upper Saddle River, NJ. All rights reserved. This material is protected under all copyright laws as they currently exist.
No portion of this material may be reproduced, in any form or by any means, without permission in writing from the publisher.

7.3 Uniform Circular Motion and Centripetal Acceleration

When an object moves with a uniform speed in a circle, the motion is called **uniform circular motion**.

Centripetal Acceleration
In a uniform circular motion, the direction of velocity of an object continually changes. This causes an acceleration ($\mathbf{a} = \Delta\mathbf{v}/\Delta t$) of the object. It can be shown that as the time interval Δt tends to zero, Δv, and hence the acceleration, points exactly toward the center of the circle. This acceleration is called the **centripetal acceleration** because it is directed toward the center of the circle.

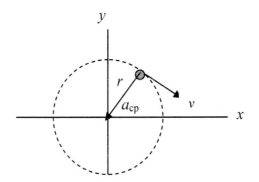

It can be shown that for an object of mass m rotating with a constant speed v in a circular path of radius r, the magnitude of the centripetal acceleration is given by

$$a_c = \frac{v^2}{r} \qquad (7.9)$$

69

2007 Pearson Education, Inc., Upper Saddle River, NJ. All rights reserved. This material is protected under all copyright laws as they currently exist. No portion of this material may be reproduced, in any form or by any means, without permission in writing from the publisher.

Since $v = r\omega$,
$$a_c = r\omega^2 \qquad (7.10)$$

The centrifuge is a rotating machine that is used to separate particles of different sizes and densities suspended in liquids, such as to separate blood components. Laboratory centrifuges typically operate at speeds to produce centripetal accelerations a thousand times more than g.

Centripetal Force

Wherever there is acceleration, there must be a net force. The force needed to cause centripetal acceleration is called **centripetal force**. From Newton's second law, the magnitude of the centripetal force is given by

$$F_c = ma_c = m\frac{v^2}{r} \qquad (7.11)$$

The centripetal force is directed toward the center of the circle.

Keep in mind that centripetal force is not a new kind of force; it is a new name for a force directed toward the center of a circle, and it can be produced in many different ways. For a ball tied to a string and that is rotating in a circle about a person's head, the tension in the string provides the centripetal force. For a car turning a corner, the force of static friction between the tires and the road is the car's centripetal force. For an orbiting satellite and for the moon orbiting the Earth, the force of gravity is the centripetal force. For a car turning on a smooth, banked road, the normal force is the centripetal force.

© 2007 Pearson Education, Inc., Upper Saddle River, NJ. All rights reserved. This material is protected under all copyright laws as they currently exist. No portion of this material may be reproduced, in any form or by any means, without permission in writing from the publisher.

7.4 Angular Acceleration

When the angular velocity of a rotating object changes, the object experiences an angular acceleration, α. The average angular acceleration is the time rate of change of angular velocity. The **average angular acceleration ($\overline{\alpha}$)** is

$$\overline{\alpha} = \frac{\Delta\omega}{\Delta t} = \frac{\omega - \omega_0}{t - t_0}$$

If $t_0 = 0$, the preceding equation gives

$$\omega = \omega_0 + \alpha t \qquad (7.12)$$

The SI unit of angular acceleration is $rad/s^2 = s^{-2}$. For circular motion with a constanr radius r, the tangential acceleration, a_t is:

$$a_t = r\alpha \qquad (7.13)$$

7.5 Newton's Law of Gravitation

Every object in the universe attracts every other object with a force called gravitation. The law is universal and is believed to apply everywhere in the universe. According to Newton's **universal law of gravitation**, the force of attraction between two objects of mass m_1 and m_2 is given by

$$F = G\frac{m_1 m_2}{r^2} \qquad (7.14)$$

where r is the distance between the masses and G is a constant called the **universal gravitational constant**. Its value is

71

2007 Pearson Education, Inc., Upper Saddle River, NJ. All rights reserved. This material is protected under all copyright laws as they currently exist. No portion of this material may be reproduced, in any form or by any means, without permission in writing from the publisher.

$$G = 6.67 \times 10^{-11} \text{ N} \cdot \text{m}^2/\text{kg}^2$$

Cavendish determined the value of G in 1798 using a sensitive balance to measure the gravitational force between separated spherical masses.

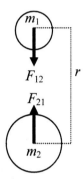

Clearly, the force of gravity decreases with distance r as $1/r^2$. This relationship is called the *inverse square law*.

The force of gravity is an action-reaction force. That is, the force (F_{12}) on m_1 is equal in magnitude but in opposite direction to the force (F_{21}) exerted on m_2.

The force of gravity between two ordinary objects on Earth is negligibly small, since G is very small. The force of gravity by Earth on a macroscopic object on or near the surface of Earth is noticeable, however.

The acceleration due to gravity, a_g, at a distance r from the center of a planet of mass M can be obtained by equating the force of gravitation on an object at that distance with ma_g.

$$ma_g = G\, \frac{mM}{r^2}$$

© 2007 Pearson Education, Inc., Upper Saddle River, NJ. All rights reserved. This material is protected under all copyright laws as they currently exist.
No portion of this material may be reproduced, in any form or by any means, without permission in writing from the publisher.

This gives

$$a_g = \frac{GM}{r^2} \qquad (7.15)$$

Thus, the acceleration due to gravity (and hence the weight of an object) decreases inversely as the square of the distance of a planet from its center. From the preceding equation, on the surface of Earth,

$$g = \frac{GM_E}{R_E^2} \qquad (7.16)$$

assuming Earth to be a homogeneous sphere of radius R_E and mass M_E. The preceding equation is used to determine the mass of Earth from the knowledge of other quantities in the expression.

At a distance h from the surface of Earth $r = R_E + h$, thus, from Equation (7.16),

$$a_g = \frac{GM}{\left(R_E + h\right)^2} \qquad (7.17)$$

The preceding expression shows how g varies with the height h from the surface of Earth.

It can be shown that the gravitational potential energy between two point objects of mass m_1 and m_2 separated by r is

$$U = -G \frac{m_1 m_2}{r} \qquad (7.18)$$

73

2007 Pearson Education, Inc., Upper Saddle River, NJ. All rights reserved. This material is protected under all copyright laws as they currently exist. No portion of this material may be reproduced, in any form or by any means, without permission in writing from the publisher.

Clearly, U approaches zero as r tends to infinity. Although it is not immediately apparent, the preceding equation is consistent with the fact that when an object changes its height by h near Earth's surface, its change in potential energy, ΔU, is equal to mgh. The gravitational potential energy at a height h from the surface of the Earth is

$$U = -G\frac{mM_E}{r + h} \qquad (7.19)$$

The total mechanical energy of a mass m_1 moving near a stationary mass m_2 is

$$E = K + U = \frac{1}{2}m_1v^2 - G\frac{m_1 m_2}{r} \qquad (7.20)$$

Gravitational potential energy is a scalar quantity. The total gravitational potential energy for a system of objects is the algebraic sum of the gravitational potential energies of each pair of objects treated separately. For three masses m_1, m_2, and m_3, the total gravitational potential energy:

$$U = -G\frac{m_1 m_2}{r_{12}} - G\frac{m_1 m_3}{r_{13}} - G\frac{m_2 m_3}{r_{23}} \qquad (7.21)$$

7.6 Kepler's Laws and Earth Satellites

Kepler formulated three laws that govern the motion of planets around the sun.

Kepler's First Law (the law of orbits)
Planets move in elliptic orbits with the sun at one focus.

© 2007 Pearson Education, Inc., Upper Saddle River, NJ. All rights reserved. This material is protected under all copyright laws as they currently exist.
No portion of this material may be reproduced, in any form or by any means, without permission in writing from the publisher.

Kepler's Second Law (the law of areas)

A line from the sun to a planet sweeps out equal areas in an equal amount of time.

As a result of the second law, a planet moves faster as it approaches the sun and moves more slowly as it goes away from the sun.

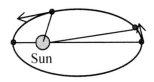

Sun

Kepler's Third Law (the law of periods)

The square of the orbital period of a planet is directly proportional to the cube of the average distance of the planet from the sun, that is, $T^2 \propto r^3$.

Kepler's laws of planetary motion can be derived easily for the special case of a planet with a circular orbit. In this case gravity provides the necessary centripetal force. Thus, for a planet of mass m_p and the sun of mass M_s, the force of gravity between them is

$$\frac{m_p v^2}{r} = \frac{G m_p M_S}{r^2}$$

and

$$v = \frac{\sqrt{GM_S}}{r}$$

Since $v = \dfrac{2\pi r}{T}$, we get from the preceding equation

75

2007 Pearson Education, Inc., Upper Saddle River, NJ. All rights reserved. This material is protected under all copyright laws as they currently exist. No portion of this material may be reproduced, in any form or by any means, without permission in writing from the publisher.

$$T^2 = \left(\frac{4\pi^2}{GM_S}\right)r^3$$

$$T^2 = Kr^3 \tag{7.22}$$

For the solar system, $K = 2.97 \times 10^{-19}$ s²/m³. T depends on the mass orbited. Using this equation one can calculate the mass orbited, knowing the period T of the object orbiting.

Earth's Satellite

The minimum speed needed to project an object to escape from the surface of a planet or moon is known as its **escape speed**. In the limit of infinite separation, the object's speed becomes zero, and since its gravitational potential energy is zero when r is infinity, its total energy is zero. If v_{esc} is the escape speed, then from the conservation of mechanical energy,

$$\frac{1}{2}mv^2_{esc} - G\frac{mM_E}{r} = 0$$

This gives

$$v_{esc} = \sqrt{\frac{2GM_E}{R_E}} \tag{7.23}$$

$$= \sqrt{2gR_E} \tag{7.24}$$

The result is about 11 km/s from the surface of the Earth. For the moon the escape speed is only 2370 m/s. Because of this low value of escape speed, atoms and molecules escaped from the atmosphere of the moon, which explains why there is currently no atmosphere on the moon.

© 2007 Pearson Education, Inc., Upper Saddle River, NJ. All rights reserved. This material is protected under all copyright laws as they currently exist.
No portion of this material may be reproduced, in any form or by any means, without permission in writing from the publisher.

Satellites are objects that orbit a planet. A tangential speed less than the escape speed is needed for a satellite in orbit. For an satellite orbiting in a circle,

$$\frac{mv^2}{r} = \frac{GmM_E}{r^2}$$

Thus,

$$v = \frac{\sqrt{GM_E}}{r} \tag{7.25}$$

Clearly, orbital speed decreases as r increases. At an altitude of 500 km, the speed of a satellite is about 7.6 km/s (that is, 4.7 mi/s).

The total energy of a satellite

$$E = K + U = \frac{1}{2}mv^2 - G\frac{mM_E}{r} \tag{7.26}$$

$$= -\frac{GmM_E}{2r} \tag{7.27}$$

using Equation (7.25).

The kinetic energy of the satellite

$$K = \frac{GmM_E}{2r} = |E| \tag{7.28}$$

An astronaut feels *apparent weightlessness* inside a satellite. This is because in this case both the astronaut and the satellite fall toward Earth at the same rate. Note that the weight measurement by a scale is the normal force N of the scale on a person standing on the scale. Thus $N = mg = w$. If the scale is placed inside an elevator that is accelerating downward, from Newton's second law
$mg - N = ma$, the apparent weight,
$w' = N = m(g - a) < mg$

77

2007 Pearson Education, Inc., Upper Saddle River, NJ. All rights reserved. This material is protected under all copyright laws as they currently exist. No portion of this material may be reproduced, in any form or by any means, without permission in writing from the publisher.

For a freely falling elevator, $a = g$, thus the apparent weight $w' = 0$. Likewise, a person inside a satellite will feel apparent weightlessness since both the astronaut and the satellite are in a state of free fall. Keep in mind, gravity is not zero at the satellite, so the true weight of the satellite is not zero.

Hints and Suggestions for Solving Problems

1. The kinematic equations for rotational motion are analogous to those for linear motion. Use the equations for rotational motion in the same way you have used kinematic equations for linear motion. In this case you will use rotational variables instead of linear ones.

2. To use $s = r\theta$, $v = r\omega$, and $a_t = r\alpha$, make sure that θ is in radians, ω is in rad/s, and α is in rad/s^2. Remember π rad $= 180°$. θ, ω, and α are all *positive for counter-clockwise rotations*.

3. For an object moving with a uniform speed, such as in the problems in Section 7.3, the centripetal force ($f_{cp} = mv^2/r$) and the centripetal acceleration ($a_{cp} = v^2/r$) are both directed toward the center of the circle.

4. The force of gravity is *always attractive and is applicable to point objects*. The force of gravity can be obtained from the expression $F = G\, m_1 m_2 / r^2$.

5. If more than one gravitational force is involved in a problem, the net force of gravity is the *vector sum* of the individual forces.

78

© 2007 Pearson Education, Inc., Upper Saddle River, NJ. All rights reserved. This material is protected under all copyright laws as they currently exist. No portion of this material may be reproduced, in any form or by any means, without permission in writing from the publisher.

6. A symmetrical object behaves as if all its mass were concentrated at its center. Use this fact to calculate the force of gravity between two extended symmetrical objects. Newton's law of gravitation will be useful for solving problems in Sections 7.5 and 7.6.

7. The acceleration due to gravity decreases with height h according to the relation, $a_g = GM/(R_E + h)^2$.

8. Use Kepler's third law, $T^2 = \left(4\pi^2/GM_S\right)r^3$ to calculate the period of rotation of a planet around the sun or the period of rotation of a satellite around Earth or any other planet, such as in problems in Section 7.6. Once you know the period, T, you can calculate the speed, v (circumference of the orbit/period).

9. Remember that gravitational potential energy is a scalar quantity that can be calculated from the expression $U = -Gm_1m_2/r$. For a system containing more than two objects, total U is the *algebraic sum* of the gravitational potential energies of each pair of objects separately.

10. Remember that the escape velocity of a planet depends on the mass and radius of the planet and is the same for all objects on the planet.

11. For an object in an accelerating elevator, its apparent weight is different from its true weight. If the elevator accelerates upward, the apparent weight is more, whereas if the elevator accelerates downward, the apparent weight is less.

2007 Pearson Education, Inc., Upper Saddle River, NJ. All rights reserved. This material is protected under all copyright laws as they currently exist.
No portion of this material may be reproduced, in any form or by any means, without permission in writing from the publisher.

CHAPTER 8
ROTATIONAL MOTION AND EQUILIBRIUM

In this chapter various aspects of rotational motion are described. Concepts of torque, angular momentum, and conservation of angular momentum are described.

Key Terms and Concepts

Torque
Translational equilibrium
Rotational equilibrium
Mechanical equilibrium
Stable equilibrium and unstable equilibrium
Moment of inertia
Rotational form of Newton's second law
Parallel-axis theorem
Rotational work and rotational power
Rotational kinetic energy
Angular momentum
Conservation of angular momentum

8.1 Rigid Bodies, Translations, and Rotations

If the distances between particles are constant in an object or in a system, it is called a **rigid body**. A rigid body may have either or both translational and rotational motions. Every particle in an object has the same instantaneous velocity, if the object has only **translational motion**. In a **rotational motion**, all the particles in an object rotate about an axis. A common example of rigid body motion that involves both translational and rotational motions is

80

© 2007 Pearson Education, Inc., Upper Saddle River, NJ. All rights reserved. This material is protected under all copyright laws as they currently exist. No portion of this material may be reproduced, in any form or by any means, without permission in writing from the publisher.

rolling motion. At each instant, a rolling object rotates about its **instantaneous axis of rotation**. When an object rolls without slipping,

$$v_{CM} = r\omega \qquad (8.1)$$

where v_{CM} is the speed of the center of mass, r is the radius of the object, and ω is the angular speed. Equation (8.1) is known as *the condition for rolling without slipping*. The preceding condition is also expressed as

$$s = r\theta \qquad (8.1a)$$

where s is the distance the center of mass moves. Also, the acceleration of the center of mass is

$$a_{CM} = r\alpha \qquad (8.1b)$$

8.2 Torque, Equilibrium, and Stability

Torque

The ability of a force to rotate an object is measured by **torque**, τ. The rate of change of rotation depends not only on the magnitude of the force but also on the perpendicular distance of its line of action from the axis of rotation, $r_\perp$. The perpendicular distance $r_\perp$ is called the **moment arm** or **lever arm**.

For a force applied at a distance r from the axis of rotation and at an angle θ with respect to the radial direction, the torque is

$$\tau = r_\perp F = r F \sin \theta \qquad (8.2)$$

2007 Pearson Education, Inc., Upper Saddle River, NJ. All rights reserved. This material is protected under all copyright laws as they currently exist. No portion of this material may be reproduced, in any form or by any means, without permission in writing from the publisher.

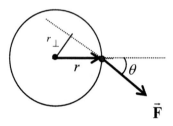

Torque is a vector quantity. The SI unit of torque is meter · newton (m · N).

Note that $\tau = 0$ when $\theta = 0$, that is, when force acts through the axis of rotation. Also, τ is maximum when $\theta = 90°$, that is force is perpendicular to r.

Torque in rotational motion is analogous to force in translational motion. As an unbalanced force changes translational motion, an unbalanced torque changes rotational motion. τ is considered to be *positive* if it causes a *counterclockwise* angular acceleration. τ is *negative* if it causes a *clockwise* angular acceleration.

Equilibrium
For an object to be in **translational equilibrium**, the sum of the forces acting on the object must be zero. For an object to be in **rotational equilibrium**, the sum of the torques acting on it must be zero. For an object to be in mechanical equilibrium it must be both in translational equilibrium and rotational equilibrium. That is,

$$\vec{F}_{net} = \sum \vec{F}_i = 0 \text{ (translational equilibrium)}$$
and   (8.3)

82

© 2007 Pearson Education, Inc., Upper Saddle River, NJ. All rights reserved. This material is protected under all copyright laws as they currently exist. No portion of this material may be reproduced, in any form or by any means, without permission in writing from the publisher.

$\tau_{net} = \Sigma\tau_i = 0$ (rotational equilibrium)

A rigid body in mechanical equilibrium is either at rest or moving at a constant linear or angular velocity. If a rigid body remains at rest, it is in **static equilibrium**.

Stability and Center of Gravity

A rigid body in equilibrium can be in stable equilibrium or unstable equilibrium. If any small displacement from equilibrium results in a restoring force or torque that tends to return the object to the original equilibrium, the object is said to be in **stable equilibrium**. For example, a ball resting inside a bowl is in stable equilibrium. On the other hand, if any small displacement from the equilibrium results in a force (and a torque) that tends to rotate the object farther away from its equilibrium position, the object is said to be in **unstable equilibrium**. For example, a ball that rests at the top of a convex surface is in unstable equilibrium.

The **center of gravity** of an object is a point at which all the weight of the object may be considered to be concentrated in representing it as a particle. An object is in stable equilibrium as long as its center of gravity after a small displacement lies above and inside its original base of support.

Rigid bodies with wide bases and low centers of gravity are most stable. This explains why a high-speed car has a wide wheelbase and its center of gravity is close to the ground.

2007 Pearson Education, Inc., Upper Saddle River, NJ. All rights reserved. This material is protected under all copyright laws as they currently exist. No portion of this material may be reproduced, in any form or by any means, without permission in writing from the publisher.

8.3 Rotational Dynamics

Moment of Inertia

A net force acting on an object produces a linear acceleration on the object. Similarly, net torque acting on an object produces an angular acceleration on the object. The magnitude of the torque acting on a particle is

$$\tau_{net} = r_\perp F_{net} = r F_\perp = r m a_\perp = mr^2 \alpha \qquad (8.4)$$

In general, for a rigid body containing masses m_i at distances r_i from the axis of rotation, the total torque acting on the rigid body can be written as

$$\Sigma \tau_{net} = (\Sigma m_i r_i^2)\alpha = I\alpha \qquad (8.5)$$

The quantity in the parentheses is called the moment of inertia, I. Thus,

$$I = \Sigma m_i r_i^2 \qquad (8.6)$$

The SI unit of I is $kg \cdot m^2$.

The preceding relationship is known as the *rotational form of Newton's second law*. Thus, just as $F = ma$ describes the dynamics of linear motion,

$$\tau_{net} = I\alpha \qquad (8.7)$$

describes the dynamics of rotational motion.

The larger the moment of inertia, I, the more resistant an object is to changes in its rotational motion, just as a larger mass m is more resistant to change in its linear motion. However, unlike the mass of a particle, the moment of

© 2007 Pearson Education, Inc., Upper Saddle River, NJ. All rights reserved. This material is protected under all copyright laws as they currently exist. No portion of this material may be reproduced, in any form or by any means, without permission in writing from the publisher.

inertia is referenced to a particular axis and can have different values for different axes.

Parallel-Axis Theorem
The axis of rotation is generally taken as the axis of symmetry passing through the center of mass. If the axis of rotation is parallel to an axis of rotation through the center of mass (such as for a rod with axis of rotation at one end of the rod), the moment of inertia of a body about such a parallel axis is given by a theorem called parallel axis theorem:

$$I = I_{CM} + Md^2 \qquad (8.8)$$

where I is the moment of inertia about an axis that is parallel to one through the center of mass, I_{CM}, and at a distance d from it, and M is the mass of the body.

Applications of Rotational Dynamics
The rotational form of Newton's second law is useful for the analyses of dynamic rotational situations such as motion of masses suspended through string from each side of a pulley of non-negligible mass, rolling motion on inclined planes, etc.

8.4 Rotational Work and Kinetic Energy

Torque acting through an angular displacement does **rotational work** in a rotational motion, just as force acting through a distance does work for linear motion. A torque τ acting through an angle θ produces work W given by

$$W = \tau\,\theta \qquad (8.9)$$

which is analogous to the expression $W = F\,d$ for linear motion.

2007 Pearson Education, Inc., Upper Saddle River, NJ. All rights reserved. This material is protected under all copyright laws as they currently exist. No portion of this material may be reproduced, in any form or by any means, without permission in writing from the publisher.

The **rotational power** is the rotational analogue of power and is given by

$$P = \tau \omega \qquad (8.10)$$

The Work–Energy Theorem and Kinetic Energy

The work–energy theorem, according to which $W = K - K_0$, applies to any motion regardless of whether the work is done by a force or by a torque.

The work done by a torque is
$$W = \tau \theta = I \alpha \theta$$
Since, $\omega^2 = \omega_0^2 + 2\alpha \theta$, this gives

$$= \frac{1}{2} I \omega^2 - \frac{1}{2} I \omega_0^2 = K - K_0, \qquad (8.11)$$

Thus, for rotational motion, kinetic energy (KE) is

$$K = \frac{1}{2} I \omega^2 \qquad (8.12)$$

The net rotational work is equal to the change in rotational kinetic energy.

Rolling motion is a combination of rotational motion and linear motion. In this case, the total kinetic energy is the sum of the kinetic energy of the center of mass and the rotational kinetic energy relative to a horizontal axis through the center of mass.

Total KE = rotational KE + translational KE

$$K = \frac{1}{2} I_{CM} \omega^2 + \frac{1}{2} M v^2_{CM} \qquad (8.13)$$

86

© 2007 Pearson Education, Inc., Upper Saddle River, NJ. All rights reserved. This material is protected under all copyright laws as they currently exist. No portion of this material may be reproduced, in any form or by any means, without permission in writing from the publisher.

8.5 Angular Momentum

An object of mass m moving with a speed v in a straight line has a linear momentum, $p = mv$. Similarly, an object of mass m moving with an angular speed ω along a circular path of radius r has an **angular momentum**, L. As torque is the product of a moment arm and a force, the angular momentum is the product of a moment arm and a linear momentum.

$$L = r_\perp p = m r_\perp v = m r_\perp^2 \omega \qquad (8.14)$$

The SI unit of angular momentum is $kg \cdot m^2/s$.

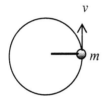

If the motion is circular, $r_\perp = r$. For a rigid body consisting of different particles, the magnitude of the total angular momentum is

$$L = (\Sigma m_i r_i^2)\, \omega = I\omega \qquad (8.15)$$

In vector notation,

$$\vec{L} = I\, \vec{\omega} \qquad (8.16)$$

The direction of $\vec{L}$ is clearly same as that of $\vec{\omega}$ as given by the right-hand rule.

Newton's second law for rotational motion can be expressed in terms of rate of change of angular momentum as

$$\tau_{net} = I\alpha = I\frac{\Delta(\omega)}{\Delta t} \quad \frac{\Delta(I\vec{\omega})}{\Delta t} = \frac{\Delta \vec{L}}{\Delta t} \qquad (8.17)$$

87

2007 Pearson Education, Inc., Upper Saddle River, NJ. All rights reserved. This material is protected under all copyright laws as they currently exist. No portion of this material may be reproduced, in any form or by any means, without permission in writing from the publisher.

Thus, the *rate of change of angular momentum is equal to the net torque*. This is the rotational analogue of

$$\vec{F}_{net} = \frac{\Delta \vec{p}}{\Delta t} \, .$$

Conservation of Angular Momentum

From Equation (8.17), if $\tau = 0$, $\Delta L = 0$. Thus **L** is constant, that is,

$$I\omega = I_0\omega_0 \tag{8.18}$$

This means that if the net external torque acting on a system is zero, the angular momentum of the system is conserved. This is called the **conservation of angular momentum**.

Conservation of angular momentum explains why a collapsing star spins faster and why a planet moves faster in its orbit as it approaches the sun.

Real-Life Angular Momentum

A person holding weights in his arms and sitting on a stool that rotates can give a popular demonstration of conservation of angular momentum. Once the stool is rotating, as the person brings his arms closer, its momentum of inertia decreases. Since the angular momentum is conserved, this causes an increase in the angular speed of the person, and he spins faster. The same principle explains why, in a hurricane, as the air rushes into the center of the storm, the rotational velocity of air increases.

When angular momentum is conserved, its direction is fixed in space. In a gyroscope a rotating wheel is universally mounted on gimbals so that it is free to rotate about any axis. When the frame moves, the wheel

88

© 2007 Pearson Education, Inc., Upper Saddle River, NJ. All rights reserved. This material is protected under all copyright laws as they currently exist. No portion of this material may be reproduced, in any form or by any means, without permission in writing from the publisher.

maintains its direction. This principle is used in the gyrocompass. A gyroscope eventually slows down because of friction causing $\vec{L}$ to tilt. As a result, its spin axis precesses about the vertical axis because of the torque given by the vertical component of the weight force. Similarly, Earth's axis of rotation precesses because of gravitational torque caused by the sun and the moon. The period of precession of Earth's axis is about 26,000 years.

Ocean tidal friction gives rise to a torque on the Earth and as a result Earth's rate of rotation is changing. The slowing rate of rotation causes an average day to be slightly longer.

Angular momentum is important in a helicopter. A helicopter has two rotors. The two rotors rotate in opposite directions to cancel each other's angular momenta.

Hints and Suggestions for Solving Problems

1. Torque is a vector quantity. The magnitude of torque can be positive, negative, or zero. Remember that the SI unit of torque is m · N (it is *not* Joule).

2. Always draw the sketch of the problem and the free-body diagram. When calculating a torque, think about the point about which the torque will be calculated. Remember that you can avoid a force that you are not interested in by taking torque about a point through which the force passes.

© 2007 Pearson Education, Inc., Upper Saddle River, NJ. All rights reserved. This material is protected under all copyright laws as they currently exist. No portion of this material may be reproduced, in any form or by any means, without permission in writing from the publisher.

3. Torque can be obtained using the expression $\tau_{net} = r_\perp F = r F \sin \theta$. If $\theta = 0$, $\tau_{net} = r F$. Use the expression for torque to solve problems in Section 8.2.

4. To solve problems relating to rotational equilibrium, such as those in Section 8.2, calculate the torque due to each force present in the system. Carefully note the *sign for each torque*. Add the individual torques, and set the result to zero ($\Sigma \tau_i = 0$) to solve for the unknown.

5. Problems relating to rotational kinetic energy and moment of inertia of an object of a system of particles or objects, such as those in Sections 8.3 and 8.4, can be solved using $K = \frac{1}{2}I\omega^2$ and $I = \Sigma m_i r_i^2$, respectively.

6. Problems relating to rotational work, such as those in Section 8.4, can be solved in the same manner as problems of work done by a force using the work–energy theorem, $W = \Delta K$. Remember that in this case $K = \frac{1}{2} I\omega^2$.

7. Read problems carefully to find out whether the net torque is zero in the problems involving angular momentum. If it is (such as in a rotational collision), apply the conservation of angular momentum ($\vec{L} = \vec{L}_0$) to solve the problem, such as those in Section 8.5.

8. Problems relating to torque and angular acceleration can be solved using $\tau_{net} = I\alpha$, which is the second law for rotational motion.

© 2007 Pearson Education, Inc., Upper Saddle River, NJ. All rights reserved. This material is protected under all copyright laws as they currently exist. No portion of this material may be reproduced, in any form or by any means, without permission in writing from the publisher.

CHAPTER 9
SOLIDS AND FLUIDS

This chapter discusses the elastic property of solids. It examines the fundamental physical principles that apply to fluids. It discusses the basic concepts needed to understand fluid flow, such as pressure, Pascal's law, buoyancy, density, Archimedes' principle, equation of continuity, Bernoulli's equation, surface tension, and viscosity.

Key Terms and Concepts

Fluid
Stress and strain
Elastic modulus
Young's modulus
Shear modulus
Bulk modulus
Pressure
Pascal
Density
Pascal's principle
Absolute pressure and gauge pressure
Buoyancy
Archimedes' principle
Flotation or sink
Equation of continuity
Bernoulli's equation
Surface tension
Viscosity
Coefficient of viscosity
Poiseuille's law

Solid, liquid, and gas are the three common phases of matter. Solids and liquids are called *condensed matter*. A

2007 Pearson Education, Inc., Upper Saddle River, NJ. All rights reserved. This material is protected under all copyright laws as they currently exist. No portion of this material may be reproduced, in any form or by any means, without permission in writing from the publisher.

fluid is a substance that can flow (liquids and gases). Intermolecular forces in liquids and gases are weaker than in solids, so their molecules can move relatively freely. Intermolecular forces in a solid are strong enough to hold its shape. Real solid bodies can be elastically deformed (usually only slightly) by external forces.

9.1 Solids and Elastic Moduli

A solid body that is slightly deformed by external forces by an applied force will return to its original shape or dimension when the force is removed. The applied force per unit of cross-sectional area of the substance is called **stress**.

$$\text{stress} = \frac{F}{A} \tag{9.1}$$

The SI unit of stress is newton per square meter (N/m^2).

Strain is a fractional change of the dimension caused by the stress.

A force applied to the ends of a rod gives *tensile stress* or *compressional stress*, depending on the direction of the force. The *tensile strain* is given by

$$\text{strain} = \frac{\text{change in length}}{\text{original length}} = \frac{\Delta L}{L_0} \tag{9.2}$$

The strain is dimensionless.

© 2007 Pearson Education, Inc., Upper Saddle River, NJ. All rights reserved. This material is protected under all copyright laws as they currently exist. No portion of this material may be reproduced, in any form or by any means, without permission in writing from the publisher.

For relatively small stresses, strain is linearly proportional to the applied stress. The constant of proportionality (which depends on the nature of the material) is called **elastic modulus**.

$$\text{elastic modulus} = \frac{\text{stress}}{\text{strain}} \qquad (9.3)$$

The SI unit of elastic modulus is newtons per square meter (N/m^2).

There are three kinds of elastic moduli, called *Young's modulus*, *shear modulus*, and *bulk modulus*, depending on whether the applied stress causes a change in length, shape, or volume, respectively.

Change in Length: Young's Modulus

If the applied stress is not too large, the plot of strain versus stress is a straight line. In this case the object comes back to its original shape when the stress is removed. This straight-line relationship is called *Hooke's law*.

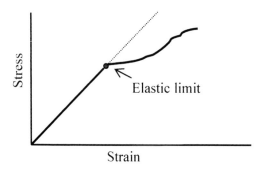

93

2007 Pearson Education, Inc., Upper Saddle River, NJ. All rights reserved. This material is protected under all copyright laws as they currently exist. No portion of this material may be reproduced, in any form or by any means, without permission in writing from the publisher.

If the amount of force F is required to produce a stretch ΔL of a solid of initial length L_0 and cross-sectional area A, then from Hooke's law:

$$\frac{F}{A} = \gamma \left(\frac{\Delta L}{L_0} \right) \quad \text{or } \gamma = \frac{F/A}{\Delta L / L_0} \qquad (9.4)$$

The constant γ in this expression is the **Young's modulus**, whose SI unit is newton per square meter (N/m^2).

Change in Shape: Shear Modulus

Equal and opposite forces applied tangentially to two surfaces of an object result in shear deformation. The applied force per unit area is called the *shearing stress*. In this case a change in shape results without any change in volume. The **shear modulus** is given by

$$S = \frac{F/A}{x/h} \approx \frac{F/A}{\phi} \qquad (9.5)$$

where $\tan \phi = x/h$, the angle ϕ is called the *shear angle*, and the ratio x/h is called the *shearing strain*.

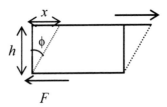

The SI unit of shear modulus is N/m^2.

© 2007 Pearson Education, Inc., Upper Saddle River, NJ. All rights reserved. This material is protected under all copyright laws as they currently exist. No portion of this material may be reproduced, in any form or by any means, without permission in writing from the publisher.

Change in Volume: Bulk Modulus

When the pressure on a solid or a fluid increases, its volume decreases. The change in pressure is equal to the *volume stress*. The *volume strain* is the ratio of the volume change to the original volume.

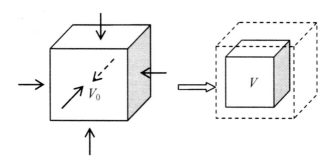

The pressure difference, ΔP, is related to the change in volume, ΔV, and initial volume, V_0, by

$$B = \frac{F/A}{-\Delta V/V_0} = \frac{\Delta p}{-\Delta V/V_0} \qquad (9.6)$$

The constant B in this expression is called the **bulk modulus** and has the units N/m^2. B is a positive quantity. The minus sign signifies that when pressure increases, volume decreases, and vice versa.

The reciprocal of bulk modulus is called the **compressibility** (k).

$$k = \frac{1}{B} \text{ (for gases)} \qquad (9.7)$$

© 2007 Pearson Education, Inc., Upper Saddle River, NJ. All rights reserved. This material is protected under all copyright laws as they currently exist. No portion of this material may be reproduced, in any form or by any means, without permission in writing from the publisher.

9.2 Fluids: Pressure and Pascal's Principle

The force, F, exerted by a fluid perpendicular to a surface divided by the area of the surface gives the **pressure**, P, of the fluid:

$$P = \frac{F}{A} \qquad (9.8a)$$

In general, $\qquad P = \frac{F \cos \theta}{A} \qquad (9.8b)$

where θ is the angle between the force F and normal to the surface area.

The SI unit of pressure is N/m^2, also called the *pascal* (Pa). In the British system, a common unit of pressure is *pounds per square inch* (lb/in^2).

The denser a material, the more mass it has in a given volume. The density, ρ, of a fluid or any material of mass M and volume V is defined as

$$\text{density} = \frac{mass}{volume}$$

$$\text{or } \rho = \frac{M}{V}$$

The SI unit of density is kg/m^3.

The density of a fluid depends on its temperature and pressure. The density of water at $4\ ^{\circ}C$ is $1000\ kg/m^3$.

Pressure and Depth

The pressure, p, of a fluid in static equilibrium increases with depth, h. This increase of pressure with depth arises because of the increasing weight of the fluid with depth and it is given by

96

© 2007 Pearson Education, Inc., Upper Saddle River, NJ. All rights reserved. This material is protected under all copyright laws as they currently exist. No portion of this material may be reproduced, in any form or by any means, without permission in writing from the publisher.

$$p = \rho g h \qquad (9.9)$$

where ρ is the density of the fluid. If the pressure on the surface of a fluid is p_0, the total pressure p at a depth h is

$$p = p_0 + \rho g h \qquad (9.10)$$

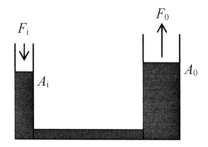

For an open container p_0 is the atmospheric pressure. Atmospheric pressure at sea level is called **atmosphere** (atm).
1 atm = 101.325 kPa = 1.01325×10^5 N/m^2 $\approx$ 14.7 lb/in^2.

Pascal's Principle
According to **Pascal's principle**, an external pressure applied to an enclosed fluid is transmitted undiminished to every point in the fluid and the enclosing walls.

97

2007 Pearson Education, Inc., Upper Saddle River, NJ. All rights reserved. This material is protected under all copyright laws as they currently exist. No portion of this material may be reproduced, in any form or by any means, without permission in writing from the publisher.

Common practical applications of this principle include hydraulic braking systems used on automobiles. A force on the brake pedal transmits a force to the wheel brake cylinder.

A hydraulic lift also works on Pascal's principle. The lift consists of two cylinders (fitted with pistons) of cross-sectional areas A_i and A_0 ($A_0 > A_i$) connected by a tube and filled with a liquid. A force F_i applied to piston 1 increases the pressure, which is transmitted through the fluid. As a result, the force F_0 on piston 2 is

$$F_0 = \left(\frac{A_0}{A_i} \right) F_i \qquad (9.11)$$

Since A_0 is greater than A_i, force is magnified in a hydraulic lift; however, the amount of work done by the two forces is the same in the absence of friction.

Pressure Measurement
Different devices can measure the pressure of fluids. One type is called an *open tube manometer*, which uses a liquid, usually mercury. This consists of a U tube, one end of which is open to the atmosphere and the other end is connected to the container of gas whose pressure is to be measured. The pressure of the gas (p) is given by

$$p = p_a + \rho g h \qquad (9.12)$$

where p_a is the atmospheric pressure. This pressure p is called the **absolute pressure**.

The **gauge pressure** p_g (such as determined by a tire gauge) is the difference between the absolute pressure and atmospheric pressure:

98

© 2007 Pearson Education, Inc., Upper Saddle River, NJ. All rights reserved. This material is protected under all copyright laws as they currently exist.
No portion of this material may be reproduced, in any form or by any means, without permission in writing from the publisher.

$$p_g = p - p_a$$

A *barometer* is used to measure atmospheric pressure. Barometers work on the principle of variation of pressure with height. If the height of a mercury column in a barometer is h, the atmospheric pressure is

$$p_a = \rho g h \qquad (9.13)$$

A *standard atmosphere* is defined as the pressure given by mercury of height 76 cm at sea level and at 0°C.

> 1 atm = 76 cm Hg = 760 mm Hg = 760 torr
> 1 mm Hg = 1 torr.

An *aneroid barometer* is also used to measure atmospheric pressure. In this case, a sensitive metal diagram on an evacuated container responds to pressure changes that are indicated on a dial. The aneroid barometer is safer and less expensive.

9.3 Buoyancy and Archimedes' Principle

When an object (a solid cube for example) is immersed in a fluid (water, for example), the upward pressure exerted by the fluid on the lower surface is more than the downward pressure on the top surface, since fluid pressure increases with depth. As a result, there is a net upward force on any object immersed in a fluid. This upward force is called the **buoyant force**.

If an object of volume V is immersed in a fluid of density ρ_f, the buoyant force is

$$F_b = \rho_f g V_f \qquad (9.14)$$

99

2007 Pearson Education, Inc., Upper Saddle River, NJ. All rights reserved. This material is protected under all copyright laws as they currently exist. No portion of this material may be reproduced, in any form or by any means, without permission in writing from the publisher.

The magnitude of this buoyant force is given by **Archimedes' principle,** which states: A body immersed wholly or partially in a fluid is buoyed up by a force equal in magnitude to the weight of the *volume of the fluid* it displaces.

Buoyancy and Density

Archimedes' principle can be applied to determine whether an object will sink or float when immersed in a fluid. It can be shown that if an object of density ρ_0 and weight w_0 is completely submerged in a fluid of density ρ_f, the buoyant force is

$$F_b = \left(\frac{\rho_f}{\rho_0}\right) w_0 \tag{9.15}$$

The object will float if the buoyant force F_b is greater than the weight w_0.

Thus, an object will float (or sink) if its average density is less (or more) than that of the liquid. An object will be in equilibrium at any depth in a fluid if the average densities of the object and the liquid are equal.

In some situations, the average density of an object is purposely varied. For example, a submarine increases its average density by flooding its tanks with seawater when it submerges. Water is pumped out from the tanks of the submarine to reduce its density in order to float.

© 2007 Pearson Education, Inc., Upper Saddle River, NJ. All rights reserved. This material is protected under all copyright laws as they currently exist. No portion of this material may be reproduced, in any form or by any means, without permission in writing from the publisher.

9.4 Fluid Dynamics and Bernoulli's Equation

Fluid motion, in general, is complex. We consider an **ideal fluid**, to have a flow that is (1) *steady*, (2) *irrotational*, (3) *nonviscous*, and (4) *incompressible*.

Steady flow means that all particles of the fluid will have the same velocity at a given point in the fluid.

Irrotational flow means a fluid element (a small volume of the fluid) has no angular velocity (that is, the flow is nonturbulent).

Nonviscous flow means viscosity is negligible.

Incompressible flow means the fluid's density is constant.

Equation of Continuity

In order that matter be conserved in the flow of a fluid, the mass of fluid passing one point per unit time must be the same as the mass of fluid passing another point. Thus, for a fluid moving from point 1 to point 2,

$$\rho_1 A_1 v_1 = \rho_2 A_2 v_2 \text{ or } \rho A v = \text{constant} \qquad (9.16)$$

This relation is known as the **equation of continuity**. Most liquids are practically incompressible. Thus, the equation of continuity for the flow of an incompressible fluid is

$$A_1 v_1 = A_2 v_2 \qquad \text{or } A v = \text{constant} \qquad (9.17)$$

It is clear from the equation of continuity that the speed of a fluid changes as the area of cross-sectional area of the tube (through which it flows) changes. Thus, for liquids,

101

007 Pearson Education, Inc., Upper Saddle River, NJ. All rights reserved. This material is protected under all copyright laws as they currently exist. No portion of this material may be reproduced, in any form or by any means, without permission in writing from the publisher.

the smaller the cross-sectional area, the faster the speed of the fluid. The preceding equation is called the **flow rate equation**.

Bernoulli's Equation

The relationship between the pressure of a fluid, its speed, and its height is known as **Bernoulli's equation**. This relation can be thought of as energy conservation per volume for a fluid.

The speed of a fluid changes as it moves from point 1 to point 2, as was seen in the equation of continuity. The pressure of a fluid also changes as it moves from point 1 to point 2. If the speed, pressure, and height of a fluid change during its motion, applying the work–energy theorem to the motion of fluid, which says that work done by the fluid is equal to the change in kinetic energy of the fluid, we get

$$p_1 + \frac{1}{2}\rho v_1^2 + \rho g y_1 = p_2 + \frac{1}{2}\rho v_2^2 + \rho g y_2$$

(9.18)

or $\qquad p + \frac{1}{2}\rho v^2 + \rho g y = \text{constant}$

© 2007 Pearson Education, Inc., Upper Saddle River, NJ. All rights reserved. This material is protected under all copyright laws as they currently exist. No portion of this material may be reproduced, in any form or by any means, without permission in writing from the publisher.

For a horizontal flow ($y_1 = y_2$), then $p + \dfrac{1}{2}\rho v^2$ = constant.

This means that if the speed of the fluid increases, its pressure decreases. Bernoulli's equation and the equation of continuity also say that if the cross-sectional area of a tube is decreased, the velocity of the fluid passing through it is increased, and the pressure is decreased.

The Bernoulli effect is partially responsible for the lift of an airplane. The shape of an airplane wing is such that air flows more rapidly over the top surface than over the lower surface. Thus, the pressure above the wing is lower than below. This pressure difference causes a net upward force, called *lift*. The greater the speed difference, the greater the lift.

*9.5 Surface Tension, Viscosity, and Poiseuille's Law

Surface Tension
The surface of water and other fluids behaves as if it were a stretched elastic membrane. This effect is called **surface tension**. Surface tension arises because the molecules in a fluid exert attractive forces (called *van der Waals' forces*) on one another. For a molecule well inside a fluid the net attractive force is zero because it experiences forces in all directions exerted by other fluid molecules. For a molecule on the surface, the net force by other fluid molecules is downward in a direction away from the surface.

Because of its surface tension, a fluid tends to pull inward on its surface, which results in a surface of minimum area.

103

2007 Pearson Education, Inc., Upper Saddle River, NJ. All rights reserved. This material is protected under all copyright laws as they currently exist. No portion of this material may be reproduced, in any form or by any means, without permission in writing from the publisher.

This explains why small water drops and soap bubbles are always spherical.

Viscosity

For fluids, viscosity is similar to the friction between solid surfaces. **Viscosity** refers to the internal friction to the flow of a fluid. Air has low viscosity, water is more viscous than air, and fluids such as honey are highly viscous. In liquids, viscosity is caused by short-range cohesive forces and in gases by collisions between molecules.

Internal friction causes fluid layers to move relative to each other in response to a shear stress. This layered motion is called *laminar* flow, which occurs at low velocities. Since there are shear stresses and shear strains in a laminar flow, viscosity is characterized by a *coefficient of viscosity*, η, which is the ratio of the shear stress to the rate of change of the shear strain.

The SI unit of coefficient of viscosity is *poiseuille* (PI). Its cgs unit is *poise* (P).

Viscosity varies with temperature. Note the viscosity grading of motor oil used in automobiles. In winter, low-viscosity oil (SAE grade 10W or 20W) should be used. On the other hand, in the summer, higher viscosity oil (SAE 30W or 40W) is used.

Poiseuille's Law

Poiseuille's law expresses the relation between the volume flow rate of a fluid, Q, and its coefficient of viscosity, η, and the pressure difference, ΔP. The law is expressed mathematically as

© 2007 Pearson Education, Inc., Upper Saddle River, NJ. All rights reserved. This material is protected under all copyright laws as they currently exist. No portion of this material may be reproduced, in any form or by any means, without permission in writing from the publisher.

$$Q = \frac{\pi r^4 \Delta P}{8\eta L} \qquad (9.19)$$

where r is the radius of the pipe and L its length. Because of the fourth power dependence of r, a small change in radius causes a large change in volume flow rate.

Hints and Suggestions for Solving Problems

1. Note that the *stress* is the ratio *force* over *area* and *strain* is the ratio *change in dimension* over the *original dimension*. *Elastic moduli* are the ratios of *stress* and *strain*. Keep these in mind as you solve the problems in Section 9.1. The unit of stress is the same as the unit of pressure (N/m^2), but strain is dimensionless.

2. The elastic modulus can be Young's modulus (γ), the bulk modulus (B), or the shear modulus (S). Stress = force (F)/area (A) or pressure (ΔP). Strain = $\Delta L/L_0$, or $\Delta V/V_0$, or x/h, depending on the problem. For shear modulus problems remember that the area, A, is the area of the surface parallel to the force.

3. Area conversions from square centimeters to square meters are very common in the problems on elasticity. Also, be sure to distinguish between diameter and radius when calculating an area or volume.

4. Be sure you are familiar with the concept of *pressure*, which is the *force per unit area*. Inside a fluid, the

105

2007 Pearson Education, Inc., Upper Saddle River, NJ. All rights reserved. This material is protected under all copyright laws as they currently exist. No portion of this material may be reproduced, in any form or by any means, without permission in writing from the publisher.

pressure P of a fluid at a point at a depth h below a point (where pressure is P_o) is $P = P_0 + \rho gh$. Do not forget to add atmospheric pressure in situations where the fluid is in an open container.

5. Pascal's principle provides mechanical advantage. Make sure you can apply the principle to static fluid situations. Problems in Section 9.2 can be solved using the concepts of pressure and Pascal's principle.

6. Make sure you understand Archimedes' principle. Think about *buoyant force* and how it is related to the *weight of the displaced fluid*. Once you are comfortable with these concepts you should be able to solve the problems in Section 9.3.

7. There are two basic equations in fluid dynamics: the *equation of continuity* and *Bernoulli's principle*. If the problem involves a change in the speed of a fluid as the area through which it flows changes, apply the equation of continuity.

8. Bernoulli's equation relates the pressure of a fluid to its height and speed. It can be used to solve problems such as those in Section 9.4. When using Bernoulli's equation, *draw a diagram*. Locate the region where the pressure, elevation, and speed are known. Locate the region where the unknown quantity needs to be determined.

© 2007 Pearson Education, Inc., Upper Saddle River, NJ. All rights reserved. This material is protected under all copyright laws as they currently exist.
No portion of this material may be reproduced, in any form or by any means, without permission in writing from the publisher.

CHAPTER 10
TEMPERATURE AND KINETIC THEORY

This chapter introduces the concept of temperature and heat as a form of energy transfer. It discusses different temperature scales, ideal gas laws, thermal expansion of substances, and the kinetic theory of gases.

Key Terms and Concepts

>Heat and Temperature
>Internal energy
>Celsius scale and Fahrenheit scale
>Boyle's law and Charles's law
>Ideal gas law
>Absolute temperature
>Absolute zero
>Kelvin temperature scale
>Linear expansion
>Thermal coefficient of linear expansion
>Thermal coefficient of area expansion
>Thermal coefficient of volume expansion
>Kinetic theory of gases
>Diffusion
>Equipartition theorem
>Degree of freedom

10.1 Temperature and Heat

Temperature is a measurement of the degree of hotness or coldness of objects. Hot objects have higher temperatures than cold objects. From the molecular point

2007 Pearson Education, Inc., Upper Saddle River, NJ. All rights reserved. This material is protected under all copyright laws as they currently exist. No portion of this material may be reproduced, in any form or by any means, without permission in writing from the publisher.

of view, temperature is related to the energy of atoms and molecules in the system.

Heat is the form of energy that is transferred between objects because of a temperature difference between them. Once the energy is transferred, the energy becomes the part of the total energy of the system, called its **internal energy**.

Objects are in thermal contact if heat can flow between them. Objects need not be in *physical* contact to be in thermal contact. Objects in *thermal contact* are said to be in *thermal equilibrium* if there is no exchange of heat between them.

10.2 The Celsius and Fahrenheit Temperature Scales

A **thermometer** measures temperature using some property of a substance that changes with temperature. The most commonly used property is thermal expansion, a change of the dimension or volume of a substance, which occurs as the temperature changes.

Most substances expand (or contract) with increasing (or decreasing) temperatures. This is called **thermal expansion**. Bimetallic strips are made of two thin strips of metals having different rates of linear expansion. The shape of a bimetallic strip is very sensitive to temperature changes. This fact is used in applications such as thermometers. As the temperature changes, the strip changes its shape, moving a needle to indicate temperature. In a liquid thermometer, the liquid in a glass tube expands into a glass stem rising in a capillary bore.

© 2007 Pearson Education, Inc., Upper Saddle River, NJ. All rights reserved. This material is protected under all copyright laws as they currently exist. No portion of this material may be reproduced, in any form or by any means, without permission in writing from the publisher.

Thermometers are calibrated so that a numerical value is assigned to a given temperature. For the definitions of any standard scale, two fixed reference points are needed. The ice point and the steam point have been chosen as the convenient fixed points for the two familiar scales: the **Fahrenheit temperature scale** and the **Celsius temperature scale**.

According to the Fahrenheit scale, the ice point and steam point are 32°F and 212°F, respectively. There are 180 equal intervals between the two reference points. According to the Celsius scale, the ice point and steam point are 0°C and 100°C, respectively. There are 100 equal intervals between the two reference points.

The Fahrenheit scale not only has a different zero than the Celsius scale, it also has a different "size" for its degree. Clearly, 180 Fahrenheit degrees change in temperature is equivalent to 100 Celsius degrees.

The conversion between degrees Fahrenheit and degrees Celsius is

$$T_F = \frac{9}{5} T_C + 32 \qquad (10.1)$$

The conversion between degrees Celsius and degrees Fahrenheit is

$$T_C = \frac{5}{9} (T_F - 32) \qquad (10.2)$$

Liquid thermometers are not adequate for accurate determination of temperatures. For sensitive temperature measurements other types of thermometers are used.

:007 Pearson Education, Inc., Upper Saddle River, NJ. All rights reserved. This material is protected under all copyright laws as they currently exist. No portion of this material may be reproduced, in any form or by any means, without permission in writing from the publisher.

10.3 Gas Laws, Absolute Temperature, and the Kelvin Temperature Scale

All gases at very low densities show the same expansion behavior. Thus, a thermometer that uses a gas gives the same temperature regardless of the gas used.

Pressure, volume, and temperature are three variables (p, V, and T) that describe the behavior of a given quantity of a gas. If the temperature T of a gas is kept constant, then

$$pV = \text{constant, or, } \quad p_1 V_1 = p_2 V_2 \qquad (10.3)$$

That is, the product of pressure and volume of the gas is constant. This relation is called *Boyle's law*.

If the pressure, p, is kept constant, then

$$\frac{V}{T} = \text{constant} \quad \text{or} \quad \frac{V_1}{T_1} = \frac{V_2}{T_2} \qquad (10.4)$$

That is, the ratio of volume to temperature is constant. This relation is known as *Charles's law*.

Low-density gases obey the preceding two laws. Combining the two laws, we get:

$$\frac{pV}{T} = \text{constant, or, } \quad \frac{p_1 V_1}{T_1} = \frac{p_2 V_2}{T_2} \qquad (10.5)$$

This relationship is known as the **ideal gas law**.

110

© 2007 Pearson Education, Inc., Upper Saddle River, NJ. All rights reserved. This material is protected under all copyright laws as they currently exist.
No portion of this material may be reproduced, in any form or by any means, without permission in writing from the publisher.

The preceding relationship can be written in the following form:

$$pV = Nk_BT \qquad (10.6)$$

where N is the number of molecules in the gas and the constant k_B is a fundamental constant of nature called the *Boltzmann constant*. $k_B = 1.38 \times 10^{-23}$ J/K. The K stands for temperature on the Kelvin scale.

Macroscopic Form of the Ideal Gas Law

The equation (10.6) can be written as

$$pV = nRT \qquad (10.7)$$

where $nR = Nk_B$, n is the number of moles (mols) of the gas and R is the *universal gas constant*. $R = 8.31$ J/(mol·K). This expression is the macroscopic form of the ideal gas law, since the quantities involved can be measured with laboratory equipment.

A **mole** of a substance is the amount of the substance that contains as many particles as there are atoms in 12 g of carbon-12. This number is called **Avogadro's number**, N_A.

$$N_A = 6.02 \times 10^{23} \text{ molecules/mol}$$

Note that Avogadro's number can be used to determine the mass of a molecule, m.

$$m = \frac{\text{formula mass (in kilograms)}}{N_A}$$

Because masses of atoms and molecules are very small, another unit is used:

$$1 \text{ atomic mass unit (u)} = 1.66054 \times 10^{-27} \text{ kg.}$$

111

2007 Pearson Education, Inc., Upper Saddle River, NJ. All rights reserved. This material is protected under all copyright laws as they currently exist. No portion of this material may be reproduced, in any form or by any means, without permission in writing from the publisher.

Absolute Zero and the Kelvin Temperature Scale

If the volume of a gas is constant, its pressure is linearly proportional to its temperature, as is seen from the ideal gas law. A *constant-volume gas thermometer* works on this principle. If the pressure of a low-density gas is plotted against its temperature, a linear graph is obtained. The temperature at which the pressure exerted by a gas is zero is called **absolute zero**. A constant-volume gas thermometer can be used to determine absolute zero. If the linear graph is extrapolated, it crosses the temperature axis at $-273.15°C$. This is absolute zero.

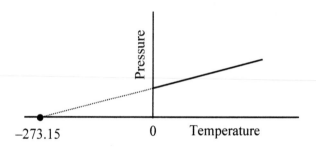

Absolute zero is the foundation of the **Kelvin temperature scale**. On this scale $-273.15°C$ is taken to be 0 K. This scale has the same degree size as the Celsius scale. The conversion between a Celsius temperature and a Kelvin temperature is

$$T_K = T_C + 273.15 \qquad (10.8)$$

The Kelvin scale is the official SI temperature scale. Keep in mind that Kelvin temperature must be used when using the ideal gas law.

112

© 2007 Pearson Education, Inc., Upper Saddle River, NJ. All rights reserved. This material is protected under all copyright laws as they currently exist. No portion of this material may be reproduced, in any form or by any means, without permission in writing from the publisher.

The Kelvin scale uses absolute zero and its second fixed point is the **triple point of water**, which represents a unique set of conditions where ice, water, and water vapor are in equilibrium.

10.4 Thermal Expansion

Most substances expand when heated and contract when cooled. Atoms in a solid are held together by bonding force, and they vibrate back and forth. Thermal expansion results from the increase in the average distance between the atoms of a substance with increasing temperature.

Thermal expansion can be linear, area, and volume expansion.

When the temperature of an object of length L_0 increases by ΔT, its length increases by ΔL, where

$$\frac{\Delta L}{L_0} = \alpha \, \Delta T \text{ or, } \Delta L = \alpha L_0 \Delta T \qquad (10.9)$$

The constant α is called the **thermal coefficient of linear expansion**. The unit of α is C^{o-1}. The value of α depends on the nature of the material.

The final length L of the substance is given by:

$$L = L_0(1 + \alpha \Delta T) \qquad (10.10)$$

Since the length of an object changes with temperature, its area and volume also change. Using Equation (10.10) we can show with first-order approximation that the final area A is related to the original area A_0 by:

113

2007 Pearson Education, Inc., Upper Saddle River, NJ. All rights reserved. This material is protected under all copyright laws as they currently exist. No portion of this material may be reproduced, in any form or by any means, without permission in writing from the publisher.

$$A = A_0(1 + 2\alpha\Delta T) \text{ or, } \frac{\Delta A}{A_0} = 2\alpha\,\Delta T \qquad (10.11)$$

Thus, the **thermal coefficient of area expansion** is 2α.

Similarly, the final volume V is given by:

$$V = V_0(1 + 3\alpha\Delta T) \text{ or, } \frac{\Delta V}{V_0} = 3\alpha\,\Delta T \qquad (10.12)$$

The thermal **coefficient of volume expansion** is 3α.

To provide space for thermal expansion because of temperature changes, bridges, sidewalks, and rail lines have gaps between different sections.

Like solids, fluids also expand with temperature. Volume expansion is meaningful only for fluids, since they do not have any fixed shape. For fluids,

$$\frac{\Delta V}{V_0} = \beta\,\Delta T \qquad (10.13)$$

where β is called the coefficient of volume expansion of fluids.

Water exhibits anomalous expansion between 0°C and 4°C. When water is heated from 0°C to 4°C, its volume decreases. This is because water molecules that were part of the open crystal structure of ice move more closely together. When water is heated above 4°C it expands normally with increasing temperature. Because of the anomalous behavior of water between 0°C and 4°C,

114

© 2007 Pearson Education, Inc., Upper Saddle River, NJ. All rights reserved. This material is protected under all copyright laws as they currently exist. No portion of this material may be reproduced, in any form or by any means, without permission in writing from the publisher.

during winter, although lakes freeze on the surface, the water below stays at 4°C, thus allowing fish and other creatures in the water to survive.

10.5 The Kinetic Theory of Gases

In an ideal gas, the molecules are viewed as point masses with relatively large distances separating them. According to the **kinetic theory of gases**, the molecules move in random directions and they undergo perfectly elastic collision with each other and with the walls of the container. Forces between molecules are of short range, and a molecule interacts with other molecules only *during* collisions.

Each collision of a molecule with the walls of a container results in a change in momentum of the molecule. The total change of momentum of the molecules per unit of time is the force exerted on the wall. The average force per unit area is the pressure exerted by the gas on the wall.

The pressure, p, of a gas is related to the volume, V, number of molecules, N, and the mass, m, of a molecule by

$$pV = \frac{1}{3} Nmv^2_{\text{rms}} \qquad (10.14)$$

where v_{rms} is called the *root-mean-square (rms)* speed. It is obtained by taking the mean of the squares of the speeds and then taking the square root of the mean.

From the preceding expression we can show that

115

2007 Pearson Education, Inc., Upper Saddle River, NJ. All rights reserved. This material is protected under all copyright laws as they currently exist. No portion of this material may be reproduced, in any form or by any means, without permission in writing from the publisher.

$$\frac{1}{2}mv_{rms}^2 = \frac{3}{2}k_BT \qquad (10.15)$$

Thus, the temperature (in the Kelvin scale) of a gas is directly proportional to the average kinetic energy of its molecules. This is the kinetic interpretation of temperature.

From Equation 10.15, at $T = 0$ K, molecular translational kinetic energy is zero. This contradicts quantum theory which states that at absolute zero the molecules will have a minimum energy called *zero-point energy*.

Internal Energy of Monatomic Gases
An ideal gas does not have potential energy, since there are no intermolecular interactions. The **internal energy**, U, of a monatomic ideal gas is the sum of the kinetic energy of all its atoms:

$$U = N\frac{1}{2}mv_{rms}^2 = \frac{3}{2}Nk_BT = \frac{3}{2}nRT \qquad (10.16)$$

Thus, the internal energy of an ideal monatomic gas is directly proportional to its absolute temperature. That is, if the absolute temperature of the gas is doubled, its internal energy is also doubled.

Diffusion
The process of transfer of molecules from a region of higher concentration to a region of lower concentration is called **diffusion**. Diffusion occurs in gas, liquids, and even, to some degree, in solids.

The rate of diffusion of a gas depends on the rms speed of its molecules. Gases can also diffuse through porous materials or permeable membranes. This process is also

© 2007 Pearson Education, Inc., Upper Saddle River, NJ. All rights reserved. This material is protected under all copyright laws as they currently exist
No portion of this material may be reproduced, in any form or by any means, without permission in writing from the publisher.

called *effusion*. Since the average kinetic energy of a molecule is proportional to its absolute temperature, lighter gas molecules (having more speed than heavier molecules at the same temperature) diffuse through the porous material faster than the heavier molecules. For example, oxygen molecules can diffuse faster than carbon dioxide molecules. By repeating the diffusion process of a mixture of two gases, eventually pure gas can be obtained.

Fluid diffusion is important to organisms. Carbon dioxide from the air diffuses into leaves in plant photosynthesis, while oxygen and water vapor diffuse out. The diffusion of liquid through a permeable membrane is called **osmosis**, a vital process in living cells.

*10.6 Kinetic Theory, Diatomic Gases, and the Equipartition Theorem

Helium, neon, argon, krypton, xenon, and radon are monatomic, and they are called *noble* or *inert* gases. Most real gases, however, are not monatomic. Air consists primarily of diatomic molecules of nitrogen (78% by volume) and oxygen (21% by volume).

The Equipartition Theorem
A monatomic gas has only translational kinetic energy. But, a diatomic molecule can rotate and vibrate in addition to moving linearly. Thus, a diatomic molecule has translational as well as rotational and vibrational kinetic energies.

A monatomic molecule has three independent ways of moving, and, thus, three independent ways of possessing energy. Each independent way a molecule can possess energy is called a **degree of freedom**. Thus, a monatomic

117

2007 Pearson Education, Inc., Upper Saddle River, NJ. All rights reserved. This material is protected under all copyright laws as they currently exist. No portion of this material may be reproduced, in any form or by any means, without permission in writing from the publisher.

molecule has three translational degrees of freedom. A diatomic gas can have vibrational kinetic and potential energies (two additional degrees of freedom). A diatomic molecule can rotate about two axes giving two degrees of freedom of rotation.

The **equipartition theorem** states how the energy of a gas is divided into its degrees of freedom. According to this theorem, the total internal energy of an ideal gas is divided equally among its various degrees of freedom, and for each degree of freedom the energy is $\frac{1}{2} k_B T$.

Since the number of degrees of freedom is three for a monatomic gas, its total energy is

$$U = 3 N \left(\frac{1}{2} k_B T\right) = \frac{3}{2} N k_B T,$$

which is the same as Equation (10.16).

The Internal Energy of a Diatomic Gas

A diatomic molecule has three translational degrees of freedom and two rotational degrees of freedom. Diatomic molecules do *not* vibrate at room temperatures. Therefore, its total internal energy is

$$U = K_{\text{trans}} + K_{\text{rot}} = 3 \left(N \frac{1}{2} k_B T\right) + 2 \left(N \frac{1}{2} k_B T\right)$$

$$= \frac{5}{2} N k_B T = \frac{5}{2} n R T \qquad (10.17)$$

118

© 2007 Pearson Education, Inc., Upper Saddle River, NJ. All rights reserved. This material is protected under all copyright laws as they currently exist. No portion of this material may be reproduced, in any form or by any means, without permission in writing from the publisher.

Hints and Suggestions for Solving Problems

1. Use Equations 10.1 and 10.2 to convert a temperature from Celsius to Fahrenheit and vice versa. Use Equation 10.8 to convert temperature from degrees Celsius to kelvins.

2. A *temperature change* in degrees Celsius is related to a *temperature change* in degrees Fahrenheit, by a factor of 5/9. A change in temperature in Celsius degrees is the same as the change in temperature in kelvins.

3. The pressure, volume, and temperature of an ideal gas are related by the equation $PV = Nk_BT = nRT$. This equation will be useful for solving problems in Section 10.3. Remember that the temperature T in this expression *must be* in kelvins. Also remember that N is the number of *molecules* and n is the number of *moles*. For problems relating to Boyle's law assume that T is constant, so that the ideal-gas equation becomes PV = constant. For problems relating to Charles's law assume that P is constant, so that the ideal gas equation becomes V/T = constant.

4. In the expression for thermal expansion the units of L_0, ΔL (or V_0, ΔV) need not be SI units as long as the units are the same, because the units cancel.

5. Use the expressions $\Delta L = \alpha L_0 \Delta T$, $\Delta A = 2\alpha A_0 \Delta T$ and $\Delta V = 3\alpha V_0 \Delta T$ to solve problems in Section 10.4.

6. Remember that a hole in a material (or an empty volume inside a container) expands at the same rate as

119

2007 Pearson Education, Inc., Upper Saddle River, NJ. All rights reserved. This material is protected under all copyright laws as they currently exist. No portion of this material may be reproduced, in any form or by any means, without permission in writing from the publisher.

if it were made of the material. To find such expansions use the coefficient of thermal expansion of the material.

7. Remember that temperature is in *kelvins* in any expression of the kinetic theory of gases involving temperature, such as the expressions for internal energy and rms molecular speed.

8. The expression $\frac{1}{2}mv_{rms}^2 = \frac{3}{2}k_BT$ will be very helpful in solving most problems dealing with the kinetic theory of gases, such as those in Section 10.5. Most other quantities such as internal energy of a gas and rms molecular speed can be obtained from this expression. This equation, together with $PV = Nk_BT$, can be used to calculate pressure.

9. Remember that for a monatomic ideal gas, internal energy $U = 3/2\ N\ k_BT$, while for a diatomic ideal gas, $U = 5/2\ N\ k_BT$. The number 3 or 5 in the expressions represents the number of degrees of freedom.

© 2007 Pearson Education, Inc., Upper Saddle River, NJ. All rights reserved. This material is protected under all copyright laws as they currently exist. No portion of this material may be reproduced, in any form or by any means, without permission in writing from the publisher.

CHAPTER 11
HEAT

This chapter discusses heat as a form of energy transfer. Specific heat, phase changes, latent heat, and different processes of heat transfer are also discussed.

Key Terms and Concepts

Heat
Kilocalorie and British thermal unit
Mechanical equivalent of heat
Specific heat
Calorimetry
Latent heat
Evaporation
Conduction
Thermal conductivity
Convection
Radiation
Stefan's law
Emissivity

11.1 Definition and Units of Heat

Heat is a form of energy that can be transferred between substances in thermal contact when there is a temperature difference.

The SI unit of heat is the joule (J). A common unit of heat is the **kilocalorie** (kcal), which is the amount of heat needed to raise the temperature of 1 kg of water from 14.5°C to 15.5°C.

2007 Pearson Education, Inc., Upper Saddle River, NJ. All rights reserved. This material is protected under all copyright laws as they currently exist. No portion of this material may be reproduced, in any form or by any means, without permission in writing from the publisher.

1 kcal = 1000 cal. In studies of nutrition a different calorie called the *Calorie* (Cal) is used. 1 Cal = 1 kcal.

Another unit of heat is the **British thermal unit** (Btu). One Btu is the amount of heat needed to raise the temperature of 1 lb of water from 63°F to 64°F. 1 Btu = 252 cal = 0.252 kcal = 1055 J.

The Mechanical Equivalent of Heat

Joule first demonstrated experimentally the equivalence between heat and mechanical work. In an experiment he observed that the gravitational potential energy of a mass used to turn paddles in water raised the temperature of the water. By measuring the mechanical work done (which is same as the potential energy of the mass) and the increase in the temperature of water, Joule found that for every 4186 J of work done, the temperature of water rose by one degree Celsius per 1 kg. Thus,

$$1 \text{ kcal} = 4186 \text{ J} = 4.186 \text{ kJ}.$$

This relationship is called the **mechanical equivalent of heat**. 1 cal = 4.186 J.

11.2 Specific Heat and Calorimetry

Specific Heats of Solids and Liquids

When heat is added to a substance (such as a solid or a liquid), its temperature usually increases. If the same amount of heat is added to two different substances, the resulting temperature change will *not* be same. This is because different substances have different molecular configurations and bonding patterns.

The amount of heat Q needed to raise the temperature of a substance is proportional to the mass, m, of the substance and the change in temperature, ΔT. Thus,

© 2007 Pearson Education, Inc., Upper Saddle River, NJ. All rights reserved. This material is protected under all copyright laws as they currently exist. No portion of this material may be reproduced, in any form or by any means, without permission in writing from the publisher.

$$Q = mc\,\Delta T \quad \text{or} \quad c = \frac{Q}{m\Delta T} \qquad (11.1)$$

The constant c is called the *specific heat capacity* or simply the **specific heat**. Specific heat is the amount of heat required to change the temperature of 1 kg of a substance by 1 $C°$. It is characteristic of a substance and is independent of mass.

The SI unit of c is J/kg $\cdot$ K = J/kg $\cdot$ $C°$

The greater the specific heat of a substance, the more heat must be added or subtracted to change its temperature. Water has a specific heat of 4186 J/kg $\cdot$ $C°$, or, 1.00 kcal /kg $\cdot$ $C°$. The constant c is always positive. Q is positive if ΔT is positive (that is, when heat is *added*). Q is negative if ΔT is negative (that is, when heat is *subtracted*).

Calorimetry
Calorimetry is the quantitative measure of heat exchange that allows determination of the specific heats of substances. The procedure involves applying the conservation of energy when two or more substances of different temperatures are brought into thermal contact and allowed to come to equilibrium. If we assume that there is no loss of heat to the surroundings, *heat lost* by the hot body equals *heat gained* by the cold body. This procedure of the determination of the specific heat is also called the method of mixtures.

Specific Heat of Gases
When heat is added to a gas, its change in volume can be significant. As a result, it is important to specify the

123

2007 Pearson Education, Inc., Upper Saddle River, NJ. All rights reserved. This material is protected under all copyright laws as they currently exist. No portion of this material may be reproduced, in any form or by any means, without permission in writing from the publisher.

conditions under which heat is added. If there is no change in volume (such as when the gas is in a rigid container), the gas does no work, and the added heat is used only to increase its internal energy. If the heat is added at a constant pressure, a part of the added heat is used to increase its internal energy while the other part is used to do some work by the gas. Thus, a gas has two specific heats: *specific heat at constant pressure*, c_p, and the *specific heat at constant volume*, c_v.

Since a gas does work when heat is added at a constant pressure, c_p is always greater than c_v.

11.3 Phase Changes and Latent Heat

Matter exists in one of three *phases*: solid, liquid, or gas. In the **solid phase** the atoms or molecules are held together by attractive bonds. When enough heat is added to a solid to break the interatomic or intermolecular bonds, the solid undergoes a phase change and becomes a liquid. The temperature at which a solid becomes a liquid is called the **melting point** of the solid. Similarly, the temperature at which a liquid becomes a solid is called the **freezing point** of the liquid.

Most solids (such as ice, metals) are called *crystalline* solids since they have orderly atomic or molecular arrangements. There are solids that are *noncrystalline* or amorphous (such as glass).

In the **liquid phase**, molecules are relatively free to move, and the addition of heat further increases the motion of molecules. When the molecules receive enough energy to become separated by large distances (compared to their diameters), the liquid changes to a **gaseous phase** or

124

© 2007 Pearson Education, Inc., Upper Saddle River, NJ. All rights reserved. This material is protected under all copyright laws as they currently exist. No portion of this material may be reproduced, in any form or by any means, without permission in writing from the publisher.

vapor phase. The constant temperature at which this phase change occurs rapidly is called the **boiling point** of the liquid. Similarly, the temperature at which a gas condenses is called the **condensation point**.

Some solids (such as dry ice) can change directly from the solid to the gaseous phase. This process is called **sublimation**.

Latent Heat

During a phase change, the temperature of a system stays constant. The added heat is used to break the molecular bonds. The amount of heat that must be added (or subtracted) from 1 kg of a substance to convert it from one phase to another is called the latent heat of the substance. Latent heat is defined as the magnitude of the heat (Q) needed for phase change *per unit mass*, or,

$$L = \frac{|Q|}{m} \tag{11.2}$$

The SI unit of L is J/kg.

The latent heat needed to melt (or fuse) a substance is called the **latent heat of fusion**, L_f. The latent heat needed to convert a liquid to a gas is called the **latent heat of vaporization**, L_v. The latent heat needed to convert a solid directly to a gas is called the *latent heat of sublimation*, L_s.

From Equation 11.2, $Q = \pm mL$ (11.3)

The heat added versus *temperature* curve for an ordinary solid (such as ice) is shown by the following curve.

2007 Pearson Education, Inc., Upper Saddle River, NJ. All rights reserved. This material is protected under all copyright laws as they currently exist. No portion of this material may be reproduced, in any form or by any means, without permission in writing from the publisher.

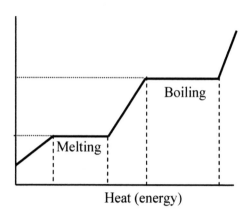

Heat (energy)

For water:

$$L_f = 3.33 \times 10^5 \text{ J/kg}$$
$$L_v = 22.6 \times 10^5 \text{ J/kg}.$$

Freezing and boiling points of water are 0°C and 100°C, respectively, at 1 atm pressure. At higher altitudes where the atmospheric pressure is less, the boiling point of water is also less. In a pressure cooker the boiling point of water is increased because of the increase in pressure inside the cooker.

Evaporation

Evaporation of a liquid can be explained by the kinetic theory of gases. The molecules in a liquid are in motion at different speeds. Evaporation occurs when molecules in the liquid phase attain speeds high enough to escape into the gas phase. The process continues without reaching equilibrium, and the molecules progressively escape into the gas phase until none is left. The escaping molecules carry energy with them. Thus, *evaporation is a cooling process* where the object from which the molecules escape cools. Evaporation depends on the temperature and the

126

© 2007 Pearson Education, Inc., Upper Saddle River, NJ. All rights reserved. This material is protected under all copyright laws as they currently exist. No portion of this material may be reproduced, in any form or by any means, without permission in writing from the publisher.

humidity. On a humid day we feel uncomfortable since the amount of evaporation from the body is less.

11.4 Heat Transfer

There are three ways heat can be transferred from one place to another: conduction, convection, and radiation.

Conduction

Conduction is the process by which heat flows through a material without any bulk motion. In this process, the atoms and molecules in a hotter section of the material vibrate or move with greater energy than those in the cooler sections. Energy is transferred by means of collisions from more energetic atoms/molecules to their less energetic neighbors.

Materials (such as metals) that conduct heat well are called **thermal conductors**. Poor conductors are called **thermal insulators**. Gases are poor thermal conductors since the molecules are farther apart. Plastic foams are good thermal insulators because they contain pockets of air.

The rate of heat flow by conduction is proportional to the surface area A and the temperature difference, ΔT, and inversely proportional to the thickness, d, of the material. Thus,

$$\frac{\Delta Q}{\Delta t} = \frac{kA\Delta T}{d} \tag{11.4}$$

where the constant k is called the **thermal conductivity** of the material. Its value depends on the nature of the material. The SI unit of k is J/(m · s · C°).

127

2007 Pearson Education, Inc., Upper Saddle River, NJ. All rights reserved. This material is protected under all copyright laws as they currently exist. No portion of this material may be reproduced, in any form or by any means, without permission in writing from the publisher.

Physics, the Construction Industry, and Energy Conservation

Equation 11.4 can be written as

$$\frac{\Delta Q}{\Delta t} = \left(\frac{1}{R_t}\right)A\Delta T \text{ where } R_t = d/k.$$ R_t is called the *thermal resistance*, which depends not only on the material properties but on the thickness of the material. More resistance is attained using thicker materials and low thermal conductivity. Insulation and building materials are classified according to "R values." In the United States the unit of R_t is $\text{ft}^2\cdot \text{h}\cdot\text{F}^\circ$ /Btu. The greater the R value the better is the insulation.

Convection

Convection is a process of heat transfer that occurs due to the bulk motion of an unevenly heated fluid. *Natural convection* occurs in liquids and gases. The sun causes greater warming near the equator than at the poles. Because of this differential heating, warm equatorial air rises, and cool polar air descends, and a global convection pattern is established. Sea breezes during the day and land breezes in the evening also arise because of convection in the air.

In *forced convection*, the medium of heat transfer is moved mechanically. Heating in homes is based mostly on forced convection. Registers or gratings in the floors or walls allow heated air to enter and cooler air to return to the heat source.

© 2007 Pearson Education, Inc., Upper Saddle River, NJ. All rights reserved. This material is protected under all copyright laws as they currently exist. No portion of this material may be reproduced, in any form or by any means, without permission in writing from the publisher.

Radiation

Radiation is a process of heat transfer in which heat flows in the form of electromagnetic waves (such as visible light and infrared). Since electromagnetic waves can propagate through a vacuum, there is no need for a physical material for radiation to occur.

We receive energy from the sun by radiation. Most of the heating effect from a burning material (such as coal) comes from the invisible **infrared radiation** emitted by the material. Heat transfer by infrared radiation is important in maintaining Earth's warmth by a process called the *greenhouse effect*. Infrared radiation is also called "heat rays."

Infrared radiation can be detected by using infrared detectors. Infrared images used in industry and medicine are called *thermograms*.

The radiated energy from an object per unit of time or radiated power, P, depends on the area, A, of the object and the *fourth power* of its absolute temperature, T. The radiated power is

$$P = \frac{\Delta Q}{\Delta t} = \sigma A e T^4 \tag{11.5}$$

The constant σ is a fundamental constant called the *Stefan–Boltzmann constant*. $\sigma = 5.67 \times 10^{-8}$ W/(m$^2 \cdot$K^4). The constant e is called the **emissivity**, which is a dimensionless number between 0 and 1 that indicates how effectively the body radiates energy.

For a perfect emitter, $e = 1$. A perfect emitter is also a perfect absorber. Such an object is called a **blackbody**. Shiny surfaces are poor absorbers since most of the incident radiations are reflected.

129

2007 Pearson Education, Inc., Upper Saddle River, NJ. All rights reserved. This material is protected under all copyright laws as they currently exist. No portion of this material may be reproduced, in any form or by any means, without permission in writing from the publisher.

If an object is placed in a surrounding whose temperature is different from the object, the object will absorb radiation from its surroundings according to the same law. As a result, the net radiated power loss of an object in surroundings of temperature T_s is

$$P_{net} = e\sigma A (T_s^4 - T^4) \qquad (11.6)$$

If T_s is less than T, there is a net loss of energy. If $T_s = T$, there is no net change of energy.

Hints and Suggestions for Solving Problems

1. In problems involving specific heat and latent heat, such as those in Sections 11.2 and 11.3, remember that *specific heat* is related to the *change in temperature in a given phase*, and *latent heat* is related to *the change of phase at a given temperature*.

2. Heat absorbed or heat given off by a substance is given by $Q = mc\Delta T$ where ΔT can be in degrees Celsius or kelvins. In calorimetric problems, such as those in Section 11.2, use this equation to calculate the heat gained by the colder objects and the heat lost by the hotter objects. Use the relation *Total heat gained = Total heat lost* to calculate the unknown in a problem such as the final temperature of a mixture, or the specific heat of a substance.

3. Heat added or removed during a phase change is given by $Q = mL$. Make sure that the unit of L is J/kg so that Q is in joules (J).

© 2007 Pearson Education, Inc., Upper Saddle River, NJ. All rights reserved. This material is protected under all copyright laws as they currently exist. No portion of this material may be reproduced, in any form or by any means, without permission in writing from the publisher.

4. For problems in heat transfer, such as those in Section 11.4, read the problem carefully to find whether the problem is related to conduction or to radiation.

5. For problems of heat flow as a result of conduction use the expression $\dfrac{\Delta Q}{\Delta t} = \dfrac{kA\Delta T}{d}$.

6. For the power radiated by a substance use the Stefan–Boltzmann law of radiation. Remember to include the factor e (emissivity) in the expression if the radiator is not a blackbody. The *net* radiated power, P, when a hot object of temperature T is placed in surroundings of temperature T_s is $P = e\sigma A (T_s^4 - T^4)$. Be careful when using this formula, as $(T_s^4 - T^4)$ is *not* equal to $(T_s - T)^4$.

2007 Pearson Education, Inc., Upper Saddle River, NJ. All rights reserved. This material is protected under all copyright laws as they currently exist. No portion of this material may be reproduced, in any form or by any means, without permission in writing from the publisher.

CHAPTER 12
THERMODYNAMICS

This chapter discusses the laws of thermodynamics.
Different thermodynamic processes and applications of the
thermodynamic concepts such as the heat engine, heat
pumps, refrigerators, and air conditioners are also
discussed.

Key Terms and Concepts

Thermodynamics
Reversible and irreversible processes
First law of thermodynamics
Isobaric process
Isometric process
Isothermal process
Adiabatic process
Second law of thermodynamics
Entropy
Isentropic process
Heat engine
Thermal efficiency
Thermal pump
Coefficient of performance
Heat pump
The Carnot cycle
Carnot efficiency
Relative efficiency
Third law of thermodynamics

Thermodynamics deals with the transfer of mechanical
work to heat and vice versa.

© 2007 Pearson Education, Inc., Upper Saddle River, NJ. All rights reserved. This material is protected under all copyright laws as they currently exist.
No portion of this material may be reproduced, in any form or by any means, without permission in writing from the publisher.

12.1 Thermodynamic Systems, States, and Processes

The term **system** in thermodynamics means a quantity of matter of interest that is separated by real or imaginary boundaries from its surroundings. A system may exchange heat with its surroundings. If heat transfer into or out of the system is not possible, the system is called a **thermally isolated system**. A **heat reservoir** is a system with unlimited heat capacity so that its temperature does not change if heat is withdrawn or added to the reservoir.

State of a System

The conditions of thermodynamic systems are expressed by **equations of state**. Such equations express mathematical relations connecting the variables, called *state variables*. For example, the state of a system of a given mass in a closed system can be specified by the state variables pressure (p), volume (V), and temperature (T). The ideal gas law $pV = Nk_BT$ is a simple equation of state for such a system.

Processes

A **process** is a change in the state or the change of the thermodynamic coordinates of a system. Such a process is usually expressed on a p-V diagram, where pressure (p) is plotted along the y-axis and the volume (V) is plotted along the x-axis. If the state of the system changes very rapidly (such as a rapid expansion) so that the system passes through intermediate nonequilibrium states, the process is called an **irreversible process**. On a p-V diagram such a process can show only the initial and final equilibrium states but not the intermediate states. On the other hand, if the change occurs very slowly so that the system passes

133

2007 Pearson Education, Inc., Upper Saddle River, NJ. All rights reserved. This material is protected under all copyright laws as they currently exist. No portion of this material may be reproduced, in any form or by any means, without permission in writing from the publisher.

through a series of intermediate equilibrium states (at least in principle) between the initial and final equilibrium states, the process is called a **reversible process**.

All real processes are irreversible to some extent; however, it is possible to have a process that closely approximates a perfectly reversible process.

12.2 The First Law of Thermodynamics

The **first law of thermodynamics** is a statement of conservation of energy applied to a thermodynamic system. In general, when heat is added to a system, its internal energy increases and work is being done by the system. If the internal energy of a system changes from an initial value U_1 to a final value U_2, and W is the work done by the system due to the addition of heat Q, then

$$Q = \Delta U + W \tag{12.1}$$

where $\Delta U = U_2 - U_1$. This is the mathematical expression of the first law of thermodynamics.

According to the sign convention:

Q is *positive*	when a system *gains* heat
Q is *negative*	when a system *loses* heat
W is *positive*	if work is done *by* a system
W is *negative*	if work is done *on* a system

When a gas expands (or contracts) by a small volume ΔV, the work done by (or on) the gas is given by $W = p\,\Delta V$. It can be shown that the work done is equal to the area under the process curve on a p-V diagram.

134

© 2007 Pearson Education, Inc., Upper Saddle River, NJ. All rights reserved. This material is protected under all copyright laws as they currently exist. No portion of this material may be reproduced, in any form or by any means, without permission in writing from the publisher.

Note that when heat is added to (or removed from) a system, work is done by (or on) the system, and the system moves from one state to another state, each having a particular internal energy, U. The internal energy, U, of a system depends on the state of the system but does not depend on the way the system changed from one state to another.

The internal energy change of a substance can be determined by calculating the heat transfer and the work done by the gas.

12.3 Thermodynamic Processes for an Ideal Gas

The first law of thermodynamics can be applied to different processes for a closed system of an ideal gas.

Isothermal Process
Isothermal process is a process for which temperature, T, remains constant. Since for an isothermal process, $pV = Nk_BT = $ constant, the process path (called *isotherm*) is a hyperbola.

Since the internal energy of an ideal gas depends only on its temperature, $\Delta U = 0$ for an ideal gas in an isothermal process. Thus, from the first law
$$Q = W \tag{12.2}$$
The magnitude of the work done is given by the area under the p-v curve and is

$$W_{\text{isothermal}} = nRT \ln\left(\frac{V_2}{V_1}\right) \tag{12.3}$$

where V_1 and V_2 are initial and final volumes, respectively.

2007 Pearson Education, Inc., Upper Saddle River, NJ. All rights reserved. This material is protected under all copyright laws as they currently exist. No portion of this material may be reproduced, in any form or by any means, without permission in writing from the publisher.

Isobaric Process
Isobaric process is a process for which pressure is constant. If the pressure, p, is kept constant, the work done by a gas as its volume increases from V_1 to V_2 is

$$W_{isobar} = p\,\Delta V = P(V_2 - V_1) \qquad (12.4)$$
$$= \text{area under the horizontal line on the}$$
$$p\text{-}V \text{ diagram.}$$

Pressure

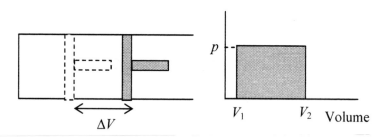

ΔV

V_1 V_2 Volume

The result that the work done is equal to the area under the curve in a p-V diagram is general. However, for a nonisobaric process the area is not a rectangle. Thus, work depends on the process and the initial and final states of the system.

From the first law, for an isobaric process

$$Q = \Delta U + p\Delta V \qquad (12.5)$$

Isometric Process
Isometric process (also called isochoric process) is a process for which volume is constant. Since there is no displacement of any of the walls of the system, no work is done ($W = 0$). Thus, from the first law

$$Q = \Delta U \qquad (12.6)$$

136

© 2007 Pearson Education, Inc., Upper Saddle River, NJ. All rights reserved. This material is protected under all copyright laws as they currently exist. No portion of this material may be reproduced, in any form or by any means, without permission in writing from the publisher.

Adiabatic Process
Adiabatic process is a process for which there is no heat transfer ($Q = 0$) between the system and its surroundings. An adiabatic process can occur when the system is thermally insulated or when the process occurs very quickly so that there is no time for heat to flow. Thus, sudden expansions or compressions are adiabatic. The path followed by the system on the *p-V* diagram in an adiabatic process is called an *adiabat*. Since $Q = 0$, from the first law

$$W = -\Delta U \qquad (12.7)$$

Thus, the internal energy of a system, and hence its temperature, decreases if work is done by the system in an adiabatic process. That is why a gas cools if it expands very quickly (adiabatic expansion). For an adiabatic process:

$$P_1 V_1^{\gamma} = P_2 V_2^{\gamma} \qquad (12.8)$$

where $\gamma = c_p/c_V$ of the gas. It can be shown that the work done by an ideal gas in an adiabatic process is

$$W_{\text{adiabatic}} = \frac{p_1 V_1 - p_2 V_2}{\gamma - 1} \qquad (12.9)$$

12.4 The Second Law of Thermodynamics and Entropy

The **second law of thermodynamics** specifies the direction in which a thermodynamic process can take place. In other words, it states that of all processes that conserve energy, only those that proceed in a certain direction can actually occur. There are several equivalent statements of the second law of thermodynamics, as mentioned next.

137

2007 Pearson Education, Inc., Upper Saddle River, NJ. All rights reserved. This material is protected under all copyright laws as they currently exist. No portion of this material may be reproduced, in any form or by any means, without permission in writing from the publisher.

The statement regarding the direction of heat flow is: *Heat does not flow spontaneously from a colder body to a hotter body.*

Another statement involves *thermal cycles*, which include several thermal processes connected in such a way that the system and its surroundings are brought back to the initial conditions at the end of the cycle. According to this statement: *In a thermal cycle, heat energy cannot be completely converted into mechanical work.*

The second law is also expressed as: *It is impossible to construct an operational perpetual motion machine.*

Entropy

Entropy is a property that determines the direction of a thermodynamic process. Like pressure, temperature, and internal energy, entropy, S, is a thermodynamic quantity. The change in entropy, ΔS, during a reversible exchange of heat Q at absolute temperature T is

$$\Delta S = \frac{Q}{T} \tag{12.10}$$

The SI unit of S is joule per kelvin (J/K).

Clearly, addition of heat to a system increases its entropy. From the molecular point of view, addition of heat increases molecular disorder in the system. Entropy is thus a measure of the disorder of a system.

As the entropy of a system increases, the energy becomes more unavailable to convert to mechanical work. Also,

138

© 2007 Pearson Education, Inc., Upper Saddle River, NJ. All rights reserved. This material is protected under all copyright laws as they currently exist. No portion of this material may be reproduced, in any form or by any means, without permission in writing from the publisher.

entropy points out the direction of time (time arrow) in which the universe moves.

It can be shown that the entropy of an isolated system increases in any natural process. This fact is used to define the second law of thermodynamics in terms of entropy as: *The entropy of the universe always increases in any natural process.*

Entropy is a state function, that is, the change in entropy depends only on the initial and final states. Since $Q = 0$, for an adiabatic process, $\Delta S = 0$. Adiabatic processes are therefore called **isentropic processes**.

12.5 Heat Engines and Thermal Pumps

A **heat engine** is a device that converts heat energy to work. It takes heat from a high-temperature source (a hot reservoir), converts a part of it to mechanical work, and returns the rest to its surroundings (a cold reservoir). Since continuous work is needed, practical heat engines usually operate in a **thermal cycle**. Steam engines and internal combustion engines (such as automobile engines) are cyclic heat engines.

It can be shown that for a cyclic heat engine, the work done per cycle is the area enclosed by the process paths in the p-V diagram composing the cycle.

Thermal Efficiency

The **thermal efficiency** ε of an engine is defined as

$$\varepsilon = \frac{\text{net work out}}{\text{heat in}} = \frac{W_{\text{net}}}{Q_{\text{in}}} \qquad (12.11)$$

139

2007 Pearson Education, Inc., Upper Saddle River, NJ. All rights reserved. This material is protected under all copyright laws as they currently exist. No portion of this material may be reproduced, in any form or by any means, without permission in writing from the publisher.

If Q_h is the heat received from a high-temperature reservoir and Q_c is the heat returned to a low-temperature reservoir,

$$Q_h - Q_c = W_{net}$$

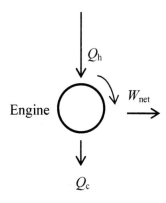

Thus,

$$\varepsilon = \frac{W_{net}}{Q_h} = 1 - \frac{Q_c}{Q_h} \tag{12.12}$$

Thermal efficiency is a dimensionless fraction and is usually expressed as a percentage. ε will be 100% when $Q_c = 0$. This is impossible according to the second law, since in that case heat will be completely converted to work.

According to the Kelvin statement of the second law of thermodynamics: *No heat engine operating in a cycle can convert its input heat completely to work.*

Thermal Pumps: Refrigerators, Air Conditioners, and Heat Pumps
A **thermal pump** is a general name for refrigerators, air conditioners, and heat pumps. These are heat engines working in reverse. They all transfer heat energy from a

140

© 2007 Pearson Education, Inc., Upper Saddle River, NJ. All rights reserved. This material is protected under all copyright laws as they currently exist. No portion of this material may be reproduced, in any form or by any means, without permission in writing from the publisher.

low-temperature reservoir to a high-temperature reservoir, when work is done on them.

A **refrigerator** uses electrical work to extract heat from the interior of a refrigerator (cold reservoir) and returns a larger amount of heat to the air in the kitchen (hot reservoir).

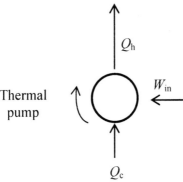

An **air conditioner** uses electrical energy to extract heat from a room (cold reservoir) and returns a larger amount of heat to the warmer air outside (hot reservoir).

The effectiveness of refrigerators and air conditioners is expressed by their **coefficient of performance** (COP), defined as the ratio of the heat extracted, Q_c, compared with the input work, W_{in}.

$$\text{COP}_{ref} = \frac{Q_c}{W_{in}} = \frac{Q_c}{Q_h - Q_c} \qquad (12.13)$$

since $Q_h = Q_c + W_{in}$.

©2007 Pearson Education, Inc., Upper Saddle River, NJ. All rights reserved. This material is protected under all copyright laws as they currently exist. No portion of this material may be reproduced, in any form or by any means, without permission in writing from the publisher.

The greater the COP, the more heat is extracted for each unit of work done. For typical refrigerators COP ranges from 3 to 5.

A **heat pump** uses electrical work to remove an amount of heat from an outdoor air (cold reservoir) and returns greater heat into the hot reservoir of air in the room.

$$COP_{hp} = \frac{Q_h}{W_{in}} = \frac{Q_h}{Q_h - Q_c} \qquad (12.14)$$

12.6 The Carnot Cycle and Ideal Heat Engines

Carnot cycle is an ideal heat engine with the maximum possible efficiency. The ideal engine absorbs heat from a constant high-temperature reservoir (T_h) and returns heat to a low-temperature reservoir (T_c). These isothermal processes are represented by two isotherms on a p-V diagram. Carnot completed the cycle with two adiabatic processes for maximum efficiency. Thus, a Carnot cycle consists of two isotherms and two adiabats and is represented by a rectangle on a T-S diagram as shown.

© 2007 Pearson Education, Inc., Upper Saddle River, NJ. All rights reserved. This material is protected under all copyright laws as they currently exist. No portion of this material may be reproduced, in any form or by any means, without permission in writing from the publisher.

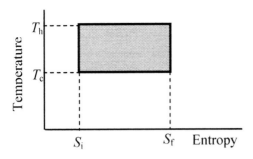

W_{net} = area of the rectangle = $(T_h - T_c)\Delta S$.

It can be shown that for a Carnot cycle,

$$\frac{Q_c}{Q_h} = \frac{T_c}{T_h}$$

Thus, the **Carnot efficiency** is given by

$$\varepsilon_C = 1 - \frac{T_c}{T_h} \qquad (12.15)$$

The efficiency, ε_C, depends only on the temperatures T_h and T_c and is always less than unity (since $T_c < T_h$). The Carnot efficiency expresses the theoretical upper limit on the thermodynamic efficiency of a heat engine.

The Third Law of Thermodynamics
It is impossible to lower the temperature of an object to absolute zero. Absolute zero can be approached closely, but cannot be attained.

143

©2007 Pearson Education, Inc., Upper Saddle River, NJ. All rights reserved. This material is protected under all copyright laws as they currently exist. No portion of this material may be reproduced, in any form or by any means, without permission in writing from the publisher.

Hints and Suggestions for Solving Problems

1. Be mindful of the signs of heat Q and work W. If *heat is added*, Q is *positive*; otherwise, it is negative. If work is done *by* the system, W is *positive*; otherwise, it is negative.

2. Heat, work, and internal energy are related by the first law $Q = \Delta U + W$. This relationship will be useful for solving problems in Sections 12.2 and 12.3.

3. To solve problems involving work done remember that for any process *the area under the p-V curve is equal to the work done, W.*

4. To calculate W using a mathematical expression for W, read the problem carefully to find out how heat is transferred, since the expression for W is different in different situations. Use Equations 12.3, 12.4, or 12.9 if the process is isothermal, isobaric, or adiabatic, respectively.

5. The efficiency of an engine can be determined using the expression $\varepsilon = W_{net}/Q_h$. $W = Q_h - Q_c$, so, $\varepsilon = 1 - Q_c/Q_h$.

6. The *coefficient of performance* (COP) of a refrigerator and an air conditioner is $COP_{ref} = Q_c/W_{in}$. Note the difference in the definition of the *coefficient of performance* for a *heat pump*: $COP_{hp} = Q_h/W_{in}$.

7. For a Carnot engine, $Q_c/Q_h = T_c/T_h$. Also, remember that all temperatures in this chapter are in kelvins.

© 2007 Pearson Education, Inc., Upper Saddle River, NJ. All rights reserved. This material is protected under all copyright laws as they currently exist.
No portion of this material may be reproduced, in any form or by any means, without permission in writing from the publisher.

8. You need to know the expressions in tips 4, 5, 6, and 7 to solve problems related to heat engines, Carnot engines, refrigerators, air conditioners, and heat pumps, such as problems in Sections 12.5 and 12.6. It is a good habit to draw an energy flow diagram in these problems.

9. The change in entropy can be calculated from the expression $\Delta S = Q/T$. The sign of S is determined by the sign of Q.

2007 Pearson Education, Inc., Upper Saddle River, NJ. All rights reserved. This material is protected under all copyright laws as they currently exist. No portion of this material may be reproduced, in any form or by any means, without permission in writing from the publisher.

CHAPTER 13
VIBRATIONS AND WAVES

This chapter describes an important kind of oscillatory motion called simple harmonic motion, which includes the motion of a mass attached to a spring and the motion of a pendulum. The chapter also describes wave motion and wave properties.

Key Terms and Concepts

Hooke's law
Simple harmonic motion
Amplitude, period, and frequency
Damped harmonic motion
Wave motion
Wave speed
Transverse and longitudinal waves
Principle of superposition
Interference
Constructive and destructive interferences
Reflection
Refraction
Dispersion
Diffraction
Standing waves
Natural frequencies
Resonance

13.1 Simple Harmonic Motion

A vibration or oscillation is a back-and-forth motion in which an object continuously repeats its path. One simple type of oscillatory motion occurs when the restoring force

© 2007 Pearson Education, Inc., Upper Saddle River, NJ. All rights reserved. This material is protected under all copyright laws as they currently exist. No portion of this material may be reproduced, in any form or by any means, without permission in writing from the publisher.

is proportional to the displacement of the object, as described by **Hooke's law**:

$$F_s = -kx \qquad (13.1)$$

where k is the spring constant. This type of oscillatory motion is called **simple harmonic motion (SHM)**. A classic example of simple harmonic motion is the motion of a mass attached to a spring.

In a simple harmonic motion, the directed distance of the object from its equilibrium position is called **displacement**, which can be positive or negative ($+x$ or $-x$). The magnitude of the maximum displacement is called **amplitude** (A). The time required to complete one cycle of motion is called the **period**, T. The SI unit of T is seconds/cycle = s.

The inverse of period is the number of cycles or oscillations per unit time and is called the **frequency**, f.

$$f = \frac{1}{T} \qquad (13.2)$$

The SI unit of f is cycles/second = 1/s = **hertz** (Hz).

Energy and Speed of a Spring–Mass System in SHM
The potential energy of a spring that is stretched or compressed by $\pm x$, is

$$U = \frac{1}{2}kx^2 \qquad (13.3)$$

The total energy of the mass on a spring is the sum of its kinetic energy, K, and its potential energy, U. Thus,

007 Pearson Education, Inc., Upper Saddle River, NJ. All rights reserved. This material is protected under all copyright laws as they currently exist. No portion of this material may be reproduced, in any form or by any means, without permission in writing from the publisher.

$$E = K + U = \frac{1}{2}mv^2 + \frac{1}{2}kx^2 \qquad (13.4)$$

Since the total energy is constant in a simple harmonic motion, when kinetic energy is a maximum (at $x = 0$), potential energy is zero. Also, when potential energy is a maximum (at the ends $x = +A$ and $x = -A$), kinetic energy is zero.

Clearly, the maximum value of U is $\frac{1}{2}kA^2$.

Therefore, $E = \frac{1}{2}kA^2 \qquad (13.5)$

Maximum kinetic energy occurs when the potential energy is zero (when the mass passes through its equilibrium position).

From the preceding two equations we can show that

$$v = \pm\sqrt{\frac{k}{m}\left(A^2 - x^2\right)} \qquad (13.6)$$

Also,

$$v_{max} = \sqrt{\frac{k}{m}}A \qquad (13.7)$$

When a spring is suspended vertically from one end, the attached mass, m, at the other end causes the spring to stretch by an amount given by
$$y_0 = mg/k$$
We can determine the spring constant of a spring experimentally from the preceding equation.

148

© 2007 Pearson Education, Inc., Upper Saddle River, NJ. All rights reserved. This material is protected under all copyright laws as they currently exist. No portion of this material may be reproduced, in any form or by any means, without permission in writing from the publisher.

13.2 Equations of Motion

If an object executes a uniform circular motion, the motion of the projection (or the shadow when illuminated from the horizontal or the vertical directions) of the object on the x- or y-axis is simple harmonic motion. We can use this connection to derive the **equation of motion** for simple harmonic motion.

For an object rotating in a circle of radius A with a constant angular speed ω, the y position of the shadow as a function of time is (if at $t = 0$, $y = 0$):

$$y = A \sin \theta = A \sin (\omega t) \qquad (13.8)$$

But if at $t = 0$, $y = A$,

$$y = A \cos \theta = A \cos (\omega t) \qquad (13.9)$$

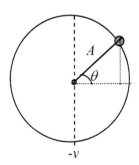

Equation 13.8 can be written as:

$$y = A \sin (2\pi f t) = A \sin \left(\frac{2\pi}{T} t \right) \qquad (13.10)$$

149

©2007 Pearson Education, Inc., Upper Saddle River, NJ. All rights reserved. This material is protected under all copyright laws as they currently exist. No portion of this material may be reproduced, in any form or by any means, without permission in writing from the publisher.

By the property of the sine and cosine functions, the position y of the mass at t and at $(t + T)$ is exactly the same. For example, at $t = 0$, $y = A$, also, at $t = T$, $x = A$ from Equation (13.9).

It can be shown that for the simple harmonic motion of a mass m attached to a spring, the period of oscillation is

$$T = 2\pi \sqrt{\frac{m}{k}} \tag{13.11}$$

$$f = \frac{1}{T} = \frac{1}{2\pi} \sqrt{\frac{k}{m}} \tag{13.12}$$

Since $\omega = 2\pi f$,

$$\omega = \sqrt{\frac{k}{m}} \tag{13.13}$$

A simple pendulum consists of a mass, m, suspended by a light string of length, L. For oscillations of small amplitude, the motion of a simple pendulum is simple harmonic. The period of oscillation is

$$T = 2\pi \sqrt{\frac{L}{g}} \tag{13.14}$$

where L is the length of the pendulum and g is the acceleration due to gravity. It is important to note that the period is independent of the mass, m, and the amplitude of oscillation, A.

© 2007 Pearson Education, Inc., Upper Saddle River, NJ. All rights reserved. This material is protected under all copyright laws as they currently exist.
No portion of this material may be reproduced, in any form or by any means, without permission in writing from the publisher.

Initial Conditions and Phase

Note that if the object starts from its equilibrium position, $y = 0$, the equation of motion is

$$y = A \sin \omega t$$

On the other hand, if the motion starts from its maximum positive displacement,

$$y = A \cos \omega t$$

In general, the initial displacement and velocity of the object executing simple harmonic motion determines the form of the equation of motion.

When two objects, both executing simple harmonic motions, can be expressed by the same equation of motion, they are said to be oscillating *in phase*. If one oscillation is ahead by 90°, their oscillations are said to be 90° *out of phase*. However, the two oscillations are *completely out of phase* if the phases differ by 180°.

Velocity and Acceleration in SHM

Velocity and acceleration of an object executing a simple harmonic motion can be easily obtained from energy and force considerations. It can be shown that for a vertical simple harmonic motion with equation $y = A \sin \omega t$, the velocity of the object is

$$v = \omega A \cos \omega t \qquad (13.15)$$

and its acceleration is

$$a = -\omega^2 A \sin \omega t = -\omega^2 y \qquad (13.16)$$

151

2007 Pearson Education, Inc., Upper Saddle River, NJ. All rights reserved. This material is protected under all copyright laws as they currently exist. No portion of this material may be reproduced, in any form or by any means, without permission in writing from the publisher.

It is important to note that in simple harmonic motion both velocity and acceleration continually change. Note that the functions for the velocity and acceleration are out of phase with that for the displacement. Also, when the magnitude of the velocity is a maximum, the acceleration is zero, and vice versa. The magnitude of the acceleration is maximum when the displacement is maximum.

Damped Harmonic Motion
In reality, energy is lost in a simple harmonic motion due to the presence of friction and air resistance. As a result, the amplitude of oscillation decreases with time. This type of motion is called **damped harmonic motion**. The time required for the motion to stop (that is, to damp out) depends on the magnitude and the type of damping force present.

In some situations damping is desirable. Damping is required for shock absorbers in an automobile and in needle indicators in instruments measuring quantities.

13.3 Wave Motion

The world is full of different waves, such as light waves, sound waves, and water waves. A **wave** motion is a transfer of disturbance from one place to another without any actual transfer of the particles of the medium. Waves carry energy. In a continuous material medium, when the particles of a medium are disturbed, the particles interact with their neighbors and transfer energy to them as they vibrate because of the restoring force acting on them. The wave propagates with well-defined speeds determined by the characteristics of the material of the medium.

© 2007 Pearson Education, Inc., Upper Saddle River, NJ. All rights reserved. This material is protected under all copyright laws as they currently exist. No portion of this material may be reproduced, in any form or by any means, without permission in writing from the publisher.

Wave Characteristics

A wave repeats itself both in space and in time. The distance over which a wave repeats is called the **wavelength** (λ) of the wave. It is actually the distance between two consecutive crests (or troughs) of a wave. The **amplitude** (A) of a wave is the magnitude of the maximum displacement. The **frequency** (f) of a wave is the number of complete waveforms that pass by a given point in one second. The reciprocal of frequency is the period (T) of a wave. It is the time for one complete waveform to pass by a given point. Since a wave travels a distance λ in time T, its speed is

$$v = \frac{\text{distance}}{\text{time}} = \frac{\lambda}{T} = \lambda f \tag{13.17}$$

Types of Waves

In a **transverse** wave, the displacement of the individual particles of the medium is perpendicular to the direction of the wave velocity. For example, if a string or rope tied to a wall at one end is pulled on the free end and moved up and down by hand, a transverse wave results. Transverse waves are also called *shear waves* since the disturbance supplies a force that tends to shear the medium. Shear waves can propagate only in solids.

In a **longitudinal** wave, the displacement of the individual particles of the medium is parallel to the direction of the wave velocity. Sound is a longitudinal wave. When a source of sound (such as a tuning fork) vibrates, it creates a series of compressions and rarefactions in the air molecules in front of the source. These compressions and rarefactions travel horizontally away from the source with the speed of sound. Longitudinal waves are sometimes

153

2007 Pearson Education, Inc., Upper Saddle River, NJ. All rights reserved. This material is protected under all copyright laws as they currently exist. No portion of this material may be reproduced, in any form or by any means, without permission in writing from the publisher.

called *compressional waves*. Longitudinal waves can travel in solids, liquids, and gases.

A water wave is a combination of transverse and longitudinal waves. In this case, at the surface the water particles move in a circular path as the wave propagates in the horizontal direction.

13.4 Wave Properties

Waves show the following properties: interference, superposition, reflection, refraction, dispersion, and diffraction.

Superposition and Interference

When two or more waves simultaneously occupy the same location, the combined waveform is given by the principle of **superposition**. According to this principle, *the resulting waveform of two or rmore interfering waves is the sum of the displacements of the individual waves at each point in the medium.* Thus, the resulting displacement is given by $y = y_1 + y_2$, where y_1 and y_2 are the individual displacements of the two waves.

As a result of superposition, when the vertical displacements of two wave pulses are in the same direction, the resulting amplitude is greater than that of the individual pulses. This situation is called **constructive interference**. On the other hand, if the individual displacements are in opposite directions, the resulting amplitude is smaller. This situation is called **destructive interference**. If the individual wave pulses are of the same width and amplitude and they meet in phase (crest to crest), the amplitude of the resulting wave pulse is twice the amplitude of the individual pulses. This is called **total**

154

© 2007 Pearson Education, Inc., Upper Saddle River, NJ. All rights reserved. This material is protected under all copyright laws as they currently exist. No portion of this material may be reproduced, in any form or by any means, without permission in writing from the publisher.

constructive interference. However, if the two pulses meet out of phase (crest to trough), the amplitude of the resulting wave pulse is zero. This is called **total destructive interference**.

Reflection, Refraction, Dispersion, and Diffraction

When a wave meets a boundary of another medium, it comes back (at least partly) to the first medium. This is called **reflection**. An echo is the reflection of sound waves.

When a wave (pulse) on a string is reflected from a fixed boundary, the reflected wave is inverted, that is, 180° out of phase. As the pulse causes the string to exert an upward force on the wall, the wall exerts an equal and opposite reaction force on the string. If the string is free to move at the boundary, there is no phase shift of the reflected wave.

When a wave enters obliquely at the boundary of two media, it changes its direction at the boundary as it moves into the second medium. This is called **refraction**. Refraction occurs because, as the wave crosses the boundary of the two media, its speed changes since the second medium has different characteristics.

A wave whose speed does *not* depend on the wavelength (or frequency) of the wave is called a *nondispersive wave*. The wave on a string is an example of a nondispersive transverse wave. A sound wave is a nondispersive longitudinal wave.

If the wave speed depends on the wavelength (or frequency), the wave shows dispersion, where waves of different component wavelengths are spread apart. White light shows dispersion when it enters water, separating its

155

©2007 Pearson Education, Inc., Upper Saddle River, NJ. All rights reserved. This material is protected under all copyright laws as they currently exist. No portion of this material may be reproduced, in any form or by any means, without permission in writing from the publisher.

different colors (wavelengths). This is the basis for the formation of a rainbow.

When a wave passes around the edge of an object or passes through a small hole, it bends. This is called **diffraction**. Because of diffraction of sound, we can hear people talking around a corner if we stand along an outside wall of a building near the corner. The eEffects of diffraction are noticeable only when the size of the diffracting object or opening is about the same size as or smaller than the wavelength of the wave. Visible light has a very small wavelength of about 10^{-6} m. That is why it does not show diffraction of light as it passes through common-sized openings.

13.5 Standing Waves and Resonance

When two identical waves moving in opposite directions are superimposed, they form a stationary pattern called a **standing wave**. It arises because of interference of the incidence waves with the reflected waves, which have same the wavelength, amplitude, and speed.

If we shake one end of a stretched rope, it forms a standing wave. Points on a standing wave where there is no displacement are called **nodes**. At these points the interfering waves cancel each other completely. Halfway between two consecutive nodes there are points where displacement is a maximum. These points are **antinodes**. At antinodes, constructive interference of the interfering waves is greatest.

A standing wave can be generated by more than one driving frequency. The frequencies at which large-amplitude standing waves are produced are called **natural**

156

© 2007 Pearson Education, Inc., Upper Saddle River, NJ. All rights reserved. This material is protected under all copyright laws as they currently exist.
No portion of this material may be reproduced, in any form or by any means, without permission in writing from the publisher.

frequencies or **resonant frequencies**. The resulting patterns are called *normal* or *resonant modes of vibration*.

The natural frequencies of a stretched string or rope can be determined from the boundary condition that there must always be a node at each end of the string (or the rope). As a result, $L = n (\lambda_n/2)$ or $\lambda_n = \dfrac{2L}{n}$ (for $n = 1, 2, 3...$) where L is the length of the string.

The natural frequencies of oscillation are given by:

$$f_n = v/\lambda_n = n\frac{v}{2L} = nf_1 \qquad (13.18)$$

The lowest frequency f_1 ($= v/2L$) is called the **fundamental frequency** or first harmonic. All other natural frequencies are integral multiples of the fundamental frequency. f_2 is the second harmonic, f_3 is the third harmonic, f_4 is the fourth harmonic, and so on.

When a string is excited in a stringed musical instrument, the resulting vibration includes the fundamental frequency and several harmonics. The number, intensity, and location of harmonics depend on how and where the string is excited. The structure of harmonics gives the musical instrument its characteristic sound.

For a stretched string, the wave speed is

$$v = \sqrt{\frac{F_T}{\mu}} \qquad (13.19)$$

© 2007 Pearson Education, Inc., Upper Saddle River, NJ. All rights reserved. This material is protected under all copyright laws as they currently exist. No portion of this material may be reproduced, in any form or by any means, without permission in writing from the publisher.

where F_T is the tension in the string and μ is the linear mass density (mass per unit length).

Thus,
$$f_n = n\frac{v}{2L} = \frac{n}{2L}\sqrt{\frac{F_T}{\mu}} = nf_1 \qquad (13.20)$$

where $n = 1, 2, 3, \ldots$

Note that the greater the μ, the lower are its natural frequencies. The greater the tension T, the greater are the natural frequencies.

When an oscillating system is driven at one of its natural frequencies, the system oscillates with very large amplitude. This condition is known as **resonance**. In this situation maximum energy transfer to the system occurs. Resonance effect is observed in a swing. If a swing is pushed with its natural frequency $(f = \frac{1}{2\pi}\sqrt{\frac{g}{L}})$ and in phase in motion, the amplitude of oscillation increases. When the frequency of the driving force matches one of the natural frequencies of oscillation in a stretched string, the maximum amount of energy is transferred to the string and a steady standing wave pattern results with large amplitude at the antinodes. Resonance plays an important role in many physical systems. Examples include a tuner in a radio or TV, atoms in a laser, and man-made structures.

© 2007 Pearson Education, Inc., Upper Saddle River, NJ. All rights reserved. This material is protected under all copyright laws as they currently exist. No portion of this material may be reproduced, in any form or by any means, without permission in writing from the publisher.

Hints and Suggestions for Solving Problems

1. Remember that simple harmonic motion occurs when the restoring force is given by $F = -kx$ (Hooke's law).

2. Do not confuse the linear frequency, f (which is in s^{-1} or Hz) with the angular frequency, ω (which is in rad/s).

3. Use the following expressions for position, velocity, and acceleration, respectively: $y = A \sin(\omega t)$ (if at $t = 0$, $y = 0$), $v = -a\omega \cos(\omega t)$, and $a = -\omega^2 A \sin(\omega t)$, where $\omega = 2\pi/T$, to solve problems illustrating the general features of a simple harmonic motion.

4. For simple harmonic motion, the basic facts to keep in mind are the definitions of spring constant ($F = kx$), the reference circle, conservation of energy, and the fact that $\omega = \sqrt{k/m}$ for a spring, and $\omega = \sqrt{g/l}$ for a pendulum. Most everything else can be quickly derived from these.

5. Energy conservation ($KE + PE = $ constant) is useful for solving many problems in this chapter. It is not necessary to remember any formulas.

©2007 Pearson Education, Inc., Upper Saddle River, NJ. All rights reserved. This material is protected under all copyright laws as they currently exist. No portion of this material may be reproduced, in any form or by any means, without permission in writing from the publisher.

CHAPTER 14
SOUND

This chapter describes sound waves, sound perception, sound intensity level, sound phenomena, Doppler effect of sound, and superposition of sounds giving rise to interference and standing waves.

Key Terms and Concepts

Sound waves
Infrasound and ultrasound
The speed of sound
Sound intensity and intensity level
Loudness
Decibel
Interference
Constructive and destructive interferences
Beats
The Doppler effect
Shock wave
Mach number
Standing waves
Quality

14.1 Sound Waves

Sound is a longitudinal wave that can travel through solids, liquids, and gases. When a tuning fork is struck, it vibrates at its fundamental frequency, and a single tone is heard. As the tuning fork vibrates, it creates alternate layers of regions with pressure higher than the normal (called *condensations*) and regions with pressure lower than the normal (called *rarefactions*). These disturbances propagate

© 2007 Pearson Education, Inc., Upper Saddle River, NJ. All rights reserved. This material is protected under all copyright laws as they currently exist. No portion of this material may be reproduced, in any form or by any means, without permission in writing from the publisher.

outward in the form of a wave. Clearly, as the sound travels it creates pressure and density variations in the air.

As the disturbances reach the ear, the eardrum is set into vibrations by the pressure variations. On the other side of the eardrum there are three bones (hammer, anvil, and stirrup) that transfer the vibrations to the inner ear, which transmits the corresponding electric signal to the brain through auditory nerves.

The **audible region** of the **sound frequency spectrum** for humans ranges from 20 Hz to 20 kHz (kilohertz).

Infrasound
Frequencies lower than 20 Hz are in the **infrasonic region**. Infrasonic waves, or *infrasound*, are generated by earthquakes and by wind and weather patterns. Elephants and cattle can hear infrasound and may give warning of earthquakes.

Ultrasound
Frequencies above 20 kHz are in the **ultrasonic region**. Ultrasonic waves or *ultrasound* can be generated by high-frequency vibrations in crystals. An ultrasound whistle is used to call dogs without disturbing people. Since ultrasound can travel for kilometers in water, it is used in sonar for ranging and detection. Bats and dolphins use ultrasound for communication or navigation. Ultrasound is used for imaging purposes in medicine (such as to view a fetus) and in lithotripsy, in which an intense beam of ultrasound is used to break kidney stones into small pieces. An ultrasonic scalpel is used in surgery that uses ultrasonic energy for precise cutting and coagulation. Ultrasonic baths are used to clean machine parts, dentures, and jewelry.

©2007 Pearson Education, Inc., Upper Saddle River, NJ. All rights reserved. This material is protected under all copyright laws as they currently exist. No portion of this material may be reproduced, in any form or by any means, without permission in writing from the publisher.

14.2 The Speed of Sound

The speed of a material wave depends on the elasticity and density of the medium through which it travels. The speeds of sound in solids and liquids are given by $v = \sqrt{Y/\rho}$ and $v = \sqrt{B/\rho}$, respectively. Here, Y is the Young's modulus, B is the bulk modulus, and ρ is the density of the medium.

Since solids are generally more elastic than liquids and liquids are more elastic than gases, the speed of sound is 2 to 4 times more in solids than liquids and about 10 to 15 times more in solids than in gases.

The speed of sound increases with temperature. The speed of sound in dry air at 0°C is

$$v = 331 \text{ m/s} = 740 \text{ mi/h}$$

At a temperature T_C (in degree Celsius) the speed of sound in air is given approximately by

$$v = (331 + 0.6\, T_C) \qquad (14.1)$$

A rough value of the speed of sound is 1/3 km/s. If the time interval between the time you observe lightning flash and hear the associated thunder is, say, 6 s, the lightning stroke is about 2 km away.

14.3 Sound Intensity and Sound Intensity Level

Sound intensity (I) is a measure of the amount of sound energy transported per unit time across a unit area.

$$\text{Intensity} = \frac{\text{power}}{\text{area}}$$

The SI unit of I is watts per square meter (W/m^2).

© 2007 Pearson Education, Inc., Upper Saddle River, NJ. All rights reserved. This material is protected under all copyright laws as they currently exist. No portion of this material may be reproduced, in any form or by any means, without permission in writing from the publisher.

For a point source of sound that sends spherical sound waves, the sound intensity at a distance R from the source is

$$I = \frac{P}{A} = \frac{P}{4\pi R^2} \qquad (14.2)$$

Clearly, the sound intensity I is inversely proportional to the square of the distance from the source. That is,

$$\frac{I_2}{I_1} = \frac{R_1^{\,2}}{R_2^{\,2}} \qquad (14.3)$$

Sound intensity is perceived by the ear as loudness. The human ear is incredibly sensitive to sound intensities. The lowest detectable sound intensity (called *threshold of hearing*) is $I_0 = 10^{-12}$ W/m^2. As the sound intensity increases, the perceived loudness increases. The intensity $I_p = 1$ W/m^2 is called the *threshold of pain*, since beyond this intensity sound is uncomfortably loud and may cause pain. Clearly, $I_p/I_0 = 10^{12}$.

Sound Intensity Level: The Bel and the Decibel

A logarithmic scale is often used to compress the large range of sound intensities and to express sound intensity level. For any intensity I, the intensity level is defined as log I/I_0. The unit of this ratio is called the **bel** (B). Thus, the intensity level is 6 B for I = 10^{-6} W/m^2.

Sound intensity level is more often expressed using a small unit called the *decibel* (dB). The **sound intensity level** or **decibel level** β is defined as

2007 Pearson Education, Inc., Upper Saddle River, NJ. All rights reserved. This material is protected under all copyright laws as they currently exist. No portion of this material may be reproduced, in any form or by any means, without permission in writing from the publisher.

$$\beta = 10 \log \left(\frac{I}{I_0} \right) \qquad (14.4)$$

where I_0 (= 10^{-12} W/m^2) is the threshold of hearing.
The intensity range is 0–120 dB in this scale.

14.4 Sound Phenomena

Reflection, Refraction, and Diffraction

When a wave reaches a boundary, it is reflected. An echo
is an evidence of sound *reflection*. Sound also refracts.
The effect of sound refraction can be experienced on a
calm summer evening when it is sometimes possible to
hear distant sounds that ordinarily would not be audible.
This happens because of *refraction* of sound as it passes
from one region to another where density is different.
Sound also shows *diffraction* as it passes around a corner.

Interference

Like any waves, sound waves *interfere* as they meet. In
constructive interference of sound, two sound waves
meet in-phase (compression to compression), and the
resulting sound is louder. In **destructive interference** of
sound, two sound waves meet out-of-phase (compression
to rarefaction), and the resulting sound is weaker.

The phase difference, $\Delta \theta$, is related to the path difference,
ΔL, between two sources by the equation

$$\Delta \theta = \frac{2\pi}{\lambda} \Delta L \qquad (14.5)$$

The following two conditions must be met for constructive
or destructive interference to occur at a point:

© 2007 Pearson Education, Inc., Upper Saddle River, NJ. All rights reserved. This material is protected under all copyright laws as they currently exist.
No portion of this material may be reproduced, in any form or by any means, without permission in writing from the publisher.

(i) $\Delta L = n\lambda$, $(n = 0, 1, 2, 3...)$ (14.6)
    for constructive interference

(ii) $\Delta L = m(\lambda/2)$ $(m = 1, 3, 5...)$ (14.7)
    for destructive interference

Destructive interference of sound is used to reduce noise levels in various situations.

When two sound waves of slightly different frequencies interfere, the resulting sound shows periodic variations of loudness in time, called **beats**. The number of times the sound loudness rises or falls in one second is called the **beat frequency (f_b)**. Beat frequency can be shown to be the difference between the two frequencies, f_1 and f_2, and is given by

$$f_b = \left| f_1 - f_2 \right| \qquad (14.8)$$

Beats are used to tune a musical instrument (such as a piano). A source of sound (such as a tuning fork) of the correct frequency is vibrated simultaneously with the instrument. The tension in the piano string is increased or decreased until the beats disappear.

14.5 The Doppler Effect

The change in the perceived frequency of sound due to the relative motion between a source of sound and an observer is called the **Doppler effect**. The observed frequency increases when the source and the observer approach each other and decreases when they move away from each other.

The frequency of sound is

165

2007 Pearson Education, Inc., Upper Saddle River, NJ. All rights reserved. This material is protected under all copyright laws as they currently exist. No portion of this material may be reproduced, in any form or by any means, without permission in writing from the publisher.

$$f = \frac{v}{\lambda}$$

where v is the speed of sound, and λ is the wavelength. When the source moves toward a stationary observer, the separation between two consecutive compressions (or rarefactions) of sound decreases due to the motion of the source. As the wavelength decreases, the observed frequency increases. The observed frequency f_o is given by

$$f_o = \left(\frac{1}{1 - \dfrac{v_s}{v}} \right) f_s \qquad (14.9)$$

where v_s and v are the speed of the source and the speed of the sound, respectively. When the source moves away from an observer,

$$f_o = \left(\frac{1}{1 + \dfrac{v_s}{v}} \right) f_s \qquad (14.10)$$

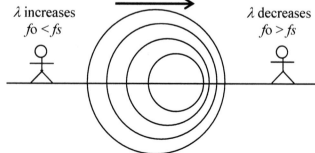

166

© 2007 Pearson Education, Inc., Upper Saddle River, NJ. All rights reserved. This material is protected under all copyright laws as they currently exist. No portion of this material may be reproduced, in any form or by any means, without permission in writing from the publisher.

In general,

$$f_o = \left(\cfrac{1}{1 \pm \cfrac{v_s}{v}} \right) f_s \tag{14.11}$$

The minus sign applies to a source *approaching* a stationary observer and the plus sign to a source *moving away* from a stationary observer.

If the observer moves toward the source with a speed v_o relative to the stationary source, the speed of sound relative to the observer is $v + v_o$. As a result, the observed frequency f_o is

$$f_o = \left(1 + \frac{v_o}{v} \right) f_s \tag{14.12}$$

When an observer moves away from a source,

$$f_o = \left(1 - \frac{v_o}{v} \right) f_s \tag{14.13}$$

In general,

$$f_o = \left(1 \pm \frac{v_o}{v} \right) f_s \tag{14.14}$$

The plus sign applies to an observer *approaching* a stationary source and the minus sign to an observer *moving away* from a stationary source.

The Doppler effect is used to determine the speed of an automobile by measuring the Doppler-shifted frequency of

167

2007 Pearson Education, Inc., Upper Saddle River, NJ. All rights reserved. This material is protected under all copyright laws as they currently exist. No portion of this material may be reproduced, in any form or by any means, without permission in writing from the publisher.

the wave reflected by the automobile. Doppler radar is used in weather forecasting to track the motion of water molecules in precipitation. In medicine, the Doppler shift is used to determine the speed of blood flow through an artery or in the heart.

The Doppler effect also occurs for light waves. Light coming from a star or galaxy moving away from Earth is shifted toward red (that is, its wavelength increases). This is called the *Doppler red shift*. Light coming from a star or galaxy approaching Earth is shifted toward blue (its wavelength decreases). This is called the *Doppler blue shift*. In astronomy, the Doppler effect is used to determine the speed and rotation of moving stars and galaxies from our knowledge of the Doppler shifts.

Sonic Booms

When a source of sound travels faster than sound (that is, travels at a supersonic speed), the overlapping waves from the source produces many points of constructive interference that form a large pressure ridge, called a *shock wave*. This is also called a *bow wave* since it is analogous to the wave produced by the bow of a boat moving through water at a speed more than the speed of water waves. When the large pressure ridge from an aircraft traveling at a supersonic speed passes over an observer on the ground, the large concentration of energy that is produced is known as a **sonic boom**.

Ideally, the sound wave produced by a supersonic aircraft is a cone-shaped shock wave. The angle between a line tangent to the spherical waves and the line along which the plane is moving is

© 2007 Pearson Education, Inc., Upper Saddle River, NJ. All rights reserved. This material is protected under all copyright laws as they currently exist. No portion of this material may be reproduced, in any form or by any means, without permission in writing from the publisher.

$$\sin \theta = \frac{v}{v_S} = \frac{1}{M} \qquad (14.15)$$

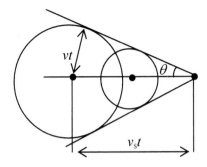

The **Mach number** (*M) is* given by the ratio

$$M = \frac{v_S}{v} = \frac{1}{\sin \theta} \qquad (14.16)$$

Clearly, a Mach number less than 1 indicates a subsonic speed, and a Mach number greater than 1 indicates a supersonic speed.

14.6 Musical Instruments and Sound Characteristics

Musical instruments are good examples of standing waves. Standing waves are formed in the vibration of an air column in wind instruments.

For an open pipe (both ends open), there will be antinodes at both ends. The natural frequencies for an open pipe are given by

169

2007 Pearson Education, Inc., Upper Saddle River, NJ. All rights reserved. This material is protected under all copyright laws as they currently exist.
No portion of this material may be reproduced, in any form or by any means, without permission in writing from the publisher.

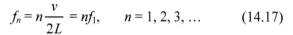

$$f_n = n\frac{v}{2L} = nf_1, \qquad n = 1, 2, 3, \ldots \qquad (14.17)$$

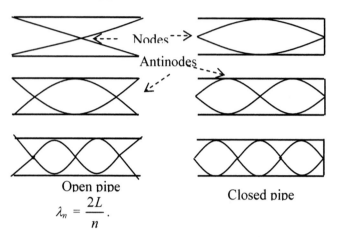

Open pipe

$$\lambda_n = \frac{2L}{n}.$$

Closed pipe

For a closed pipe (one end closed), there must be a node at the closed end and an antinode at the open end. The natural frequencies for a closed pipe are given by

$$f_n = m\frac{v}{4L} = mf_1, \qquad m = 1, 3, 5, \ldots \qquad (14.18)$$

The natural frequencies clearly depend on the length of the pipe. This principle is used in wind and brass instruments. The player varies the effective length of the pipe either by opening or closing holes or with the help of slides or valves (brass instruments).

Physically, a sound wave is characterized by its intensity, frequency, and waveform (harmonics). The corresponding perceived quantities are loudness, pitch, and quality (or timbre). However, note that the physical properties are objective and can be measured, whereas the sensory

170

© 2007 Pearson Education, Inc., Upper Saddle River, NJ. All rights reserved. This material is protected under all copyright laws as they currently exist. No portion of this material may be reproduced, in any form or by any means, without permission in writing from the publisher.

effects are subjective and can vary from person to person. Thus, frequency and pitch are not the same; there is an objective-subjective difference between them.

The curves showing intensity-frequency combinations that a person with average hearing judges to be equally loud are called equal *loudness contours* (or Fletcher–Munson curves). The human ear is most sensitive around 4000 Hz and 12000 Hz (these correspond to the first two resonant frequencies of the human ear canal). As a result, the Fletcher–Munson curve shows two dips at these frequencies.

The **quality** of a tone depends on the waveform, that is, the number and relative intensities of the harmonics. It enables us to distinguish two otherwise identical sounds coming from two different musical instruments.

Hints and Suggestions for Solving Problems

1. Note that the equation of a wave, $v = f\lambda$, is applicable to any wave, including a sound wave. You need to use this equation in many problems in this chapter.

2. Make sure you understand the difference between the sound intensity I and the sound intensity level β. They are related by $\beta = 10 \log (I/I_0)$. I = power/area. $I_0 = 10^{-12}$ W/m^2, is the threshold intensity. Doubling sound intensity I increases sound intensity level β by 3 dB. Review logarithms when solving problems on sound intensity level, such as those in Section 14.3.

2007 Pearson Education, Inc., Upper Saddle River, NJ. All rights reserved. This material is protected under all copyright laws as they currently exist. No portion of this material may be reproduced, in any form or by any means, without permission in writing from the publisher.

3. To calculate the change in frequency due to the relative motion of the source and the observer use the expression for the Doppler effect, such as for some of the problems in Section 14.5. If you are confused by the sign of any terms in the expression for the Doppler effect, remember that when the source and the observer approach each other, the apparent frequency increases. If they move away from each other, the apparent frequency decreases.

4. For *constructive interference* of two sounds, the path difference between the two sources is an *integral multiple of* λ. For *destructive interference* the path difference is an *odd multiple of* $\lambda/2$. Use this concept to solve problems on interference, such as those in Section 14.5.

5. Standing wave modes for the vibration of an air column in open and closed pipes have the same basic form, that is, the *frequency* is equal to the *velocity divided by wavelength*. It is *not* necessary to remember any formula for the modes of vibration. Relate the wavelength to the length L of the air column or the string. Remember that the distance between two consecutive nodes or antinodes is $\lambda/2$, whereas the distance between a node and the nearest antinode is $\lambda/4$. Use this concept to determine the frequencies of the different modes of vibration.

6. Remember that the *beat frequency* of two sounds is the *magnitude* of the difference between the two source frequencies and thus is always positive.

7. Keep in mind that for supersonic speed, the Mach number, $M = v_s/v$ is greater than 1.

172

© 2007 Pearson Education, Inc., Upper Saddle River, NJ. All rights reserved. This material is protected under all copyright laws as they currently exist. No portion of this material may be reproduced, in any form or by any means, without permission in writing from the publisher.

CHAPTER 15
ELECTRIC CHARGE, FORCES, AND FIELDS

This chapter discusses electric charge, the force of interaction between electric charges, electric field, and Gauss's law for electric field.

Key Terms and Concepts

Electric charge
Law of charges
Coulomb
Conservation of charge
Electrostatic charging
Charging by induction
Polarization
Coulomb's law
Electric field
Electric lines of force
Gauss's law

15.1 Electric Charge

Electric charge is a fundamental property of matter. It is fundamentally associated with the subatomic particles, the electron and the proton. Atoms contain an equal number of electrons and protons and are electrically neutral. Protons carry a positive charge and electrons carry an equal amount of negative charge. Charges exert forces to other charges. The direction of the electric forces is given by the **law of charges** or the **charge–force law**:
Like charges repel each other, and unlike charges attract each other.

173

2007 Pearson Education, Inc., Upper Saddle River, NJ. All rights reserved. This material is protected under all copyright laws as they currently exist. No portion of this material may be reproduced, in any form or by any means, without permission in writing from the publisher.

The charge of an electron is $e = -1.60 \times 10^{-19}$ C. The charge, e, of an electron is the smallest charge that has been observed in nature and is taken to be the fundamental unit of charge. An object is said to have **net charge** if it has an excess of protons or electrons. If q is the net charge of an object, then

$$q = \pm ne \qquad (15.1)$$

The plus sign or a minus sign indicate whether the object has a deficiency or an excess of electrons, respectively. The SI unit of electric charge is the **coulomb** (C). Usually charges are expressed in *microcoulombs* (μC), *nanocoulombs* (nC), and *picocoulombs* (pC).

When electrons are transferred from one object to another, the object that gains electrons will have a net negative charge, and the object that loses electrons will have a net positive charge. Charge is neither created nor destroyed in any such transfer. The principle of the **conservation of charge** states: *The net charge of an isolated system remains constant.* Since the universe as a whole is an isolated system, the total charge of the universe is constant.

15.2 Electrostatic Charging

Substances, such as metals, that allow electric charges to move easily within them are called **conductors.** The substances such as glass and rubber that do not allow electric charges to move easily within them are called **insulators**.

© 2007 Pearson Education, Inc., Upper Saddle River, NJ. All rights reserved. This material is protected under all copyright laws as they currently exist. No portion of this material may be reproduced, in any form or by any means, without permission in writing from the publisher.

In conductors, the electrons in the outermost orbits, called *valence* electrons, are loosely bound. These electrons can be easily removed from the atom, and they can move freely throughout the conductor. In insulators the valence electrons are bound to the atoms and cannot easily move within the material.

Substances, such as silicon or germanium, whose properties are intermediate between those of insulators and conductors, are called **semiconductors**. Adding certain types of atomic impurities in varying concentrations can control the conductivity of semiconductors. Semiconductors form the basis of solid-state circuits and have become the backbone of the computer industry.

An *electroscope* consists of a metal rod with a metal bulb at one end and a pair of hanging metal foil leaves at the other end. When a charged object is brought close to the bulb, the electrons in the bulb are either attracted or repelled (depending on the nature of the charge on the object). Since the electrons are conducted to or from the leaves, the leaves move apart because of the repulsive force between the same net charges on the leaves. Electroscopes demonstrate the characteristics of electric charge.

Electrostatic charging is a process in which an insulator receives a net electric charge.

Charging by Friction
When one object is rubbed with another, electrons are transferred from one substance to the other. For example, when an amber rod is rubbed with fur, the rod will acquire a net negative charge due to the transfer of electrons from the fur to the rod. This is called **charging by friction**. A

2007 Pearson Education, Inc., Upper Saddle River, NJ. All rights reserved. This material is protected under all copyright laws as they currently exist. No portion of this material may be reproduced, in any form or by any means, without permission in writing from the publisher.

glass rod rubbed with silk will give the rod a net positive charge.

On a dry day after we walk on carpet, sometimes we get a spark as we reach for a metal object such as a doorknob. This happens because the charges that we pick up by friction from the carpet flow from our hand to the doorknob, ionizing the air molecules between the hand and the doorknob.

Charging by Conduction (contact)

When a charged rod is brought close and then touched to the bulb of an electroscope, charge is transferred to the electroscope from the rod. In this case, the electroscope has been **charged by contact** or by **conduction**. When a rod of the same charge is brought near the bulb, the leaves separate further. But if an oppositely charged rod is brought close, the leaves collapse.

Charging by Induction

A conductor can be charged without being in direct contact with a charged body. This type of charging is called **charging by induction**. Touching the bulb with a finger will ground the electroscope in the preceding example. When the finger is removed, the electroscope has a net charge opposite that of the rod. This is an example of charging by induction.

Charge Separation by Polarization

Charged objects can attract small neutral objects by **polarization.** In this case, a net opposite charge (called a polarization charge) develops on the surface of the neutral object near the charged object. The polarization charges and the charges on the charged object attract each other. For example, when balloons are charged by friction and

176

© 2007 Pearson Education, Inc., Upper Saddle River, NJ. All rights reserved. This material is protected under all copyright laws as they currently exist. No portion of this material may be reproduced, in any form or by any means, without permission in writing from the publisher.

placed in contact with a wall, an opposite polarization charge is induced on the wall's surface, to which the balloons stick by force of attraction. Some molecules (such as those of water) are polar, that is, they have separated regions of positive and negative charge.

15.3 Electric Force

The magnitude of the electric force for two point charges, q_1 and q_2, is proportional to the product of the charges and inversely proportional to the square of the distance, r, between them.

$$F = k \frac{q_1 q_2}{r^2} \quad \text{(point charges only)} \quad (15.2)$$

where the constant $k = 8.988 \times 10^9 \ \text{N·m}^2/\text{C}^2$
$$\approx 9.00 \times 10^9 \ \text{N·m}^2/\text{C}^2$$

The preceding equation is called **Coulomb's law**. Coulomb's law and Newton's law of gravitation look similar. Note that the force of gravity is always attractive, whereas an electric force can be attractive or repulsive.

The electrostatic forces on two point charges are equal and opposite. The net electric force on a charge due to two or more charges is the vector sum of the electric forces due to the other charges.

15.4 Electric Field

Conceptually, an *electric field* surrounds every arrangement of charges. The electric field is a *vector field*. The strength of the **electric field** is given by the *force per unit charge*. If a small positive test charge, q_+, experiences

2007 Pearson Education, Inc., Upper Saddle River, NJ. All rights reserved. This material is protected under all copyright laws as they currently exist. No portion of this material may be reproduced, in any form or by any means, without permission in writing from the publisher.

a force $\vec{F}_{on\ q+}$ at a given location, the electric field $\vec{E}$ at the location is

$$\vec{E} = \frac{\vec{F}_{on\ q+}}{q_+} \qquad (15.3)$$

The SI unit of electric field is newton/coulomb (N/C). The electric field $\vec{E}$ is in the same direction as the force $\vec{F}_{on\ q+}$ for positive charges but in the opposite direction for negative charges.

The magnitude of the electric field due to a point charge q at a distance r is

$$E = k\frac{q}{r^2} \qquad (15.4)$$

The total or net electric field at a point due to more than one charge is the vector sum of the electric fields due to the individual charges. This is called the **superposition principle**.

Electric Lines of Force

The *electric lines of force* or **electric field lines** are pictorial representations of an electric field. The following statements hold for the electric lines of force and for sketching and interpreting electric field lines:

1. The direction of the electric field is tangential to the lines of force.

2. The electric field lines are always directed away from positive charges and toward negative charges.

178

© 2007 Pearson Education, Inc., Upper Saddle River, NJ. All rights reserved. This material is protected under all copyright laws as they currently exist. No portion of this material may be reproduced, in any form or by any means, without permission in writing from the publisher.

3. If the lines of force are closer together, the electric field is stronger.

4. The number of lines leaving or entering a charge is proportional to the magnitude of the charge.

5. No two lines of force intersect.

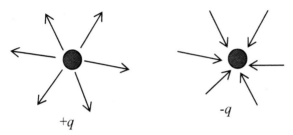

$+q$ $-q$

Electric lines of force due to point charges

Two equal and opposite charges separated by a small distance constitute an **electric dipole**. Although the net charge of a dipole is zero, the electric field due to a dipole is not zero because of the charge separation. Dipoles are important since they give a model for permanently polarized molecules, such as water molecules.
For a positively charged single horizontal plate, the net electric field is directed up (perpendicular to the surface). This is because the horizontal components of the electric fields from different locations of the plate cancel out.

Two metal plates with equal and opposite charges placed parallel to each other and separated by a small distance constitute a parallel plate capacitor.

2007 Pearson Education, Inc., Upper Saddle River, NJ. All rights reserved. This material is protected under all copyright laws as they currently exist. No portion of this material may be reproduced, in any form or by any means, without permission in writing from the publisher.

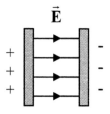

The electric field is uniform at any point between the plates and is directed perpendicularly from the positively charged plate to the negatively charged plate. The magnitude of the inside electric field E is given by

$$E = \frac{4\pi kQ}{A} \qquad (between\ the\ plates) \qquad (15.5)$$

The field is zero outside the plates.

Parallel plate capacitors are very common in electronic applications. Formation of cloud-to-ground lightning strokes can be modeled as closely spaced parallel plates to estimate the charge on the clouds.

15.5 Conductors and Electric Fields

Charges can move freely in a conductor and must experience no net electric force. *The electric field is zero everywhere inside a conductor.* As a result, *any excess charge on a conductor resides on the exterior surface of the conductor.*

The electric field at the surface of a charged conductor is perpendicular to the surface. If this were not the case, there would be a component of the field parallel to the

180

© 2007 Pearson Education, Inc., Upper Saddle River, NJ. All rights reserved. This material is protected under all copyright laws as they currently exist.
No portion of this material may be reproduced, in any form or by any means, without permission in writing from the publisher.

surface, which would move the electrons until the parallel component vanished. For a conductor of irregular shape, *excess charge tends to accumulate at sharp points or locations of greatest curvature.* As a result, the electric field is strongest at these locations.

If there is a large concentration of charge on a conductor with a sharp point, the electric field above the point may be strong enough to ionize air molecules. The resulting free electrons may also cause secondary ionizations in air molecules, resulting a visible spark. More charge can be placed on a conducting sphere before a spark can occur.

*15.6 Gauss's Law for Electric Fields: A Qualitative Approach

Gauss's law is a fundamental law of electric fields that can be used to determine the electric fields in situations of high symmetry. The law relates the number of electric field lines passing through a surface to the total charges enclosed by the surface. According to Gauss's law:
The net number of electric field lines passing through an imaginary closed surface (called a Gaussian surface) is proportional to the amount of net charge enclosed by the surface.

A net outward-pointing electric field indicates the presence of a net positive charge inside the surface (called the "source" of the electric field). On the other hand, a net inward-pointing electric field indicates the presence of a net negative charge inside (called the "sink").

Since the electric field E inside a conductor is zero, no electric field lines pass through an imaginary surface just inside the conductor having the same shape as the

181

2007 Pearson Education, Inc., Upper Saddle River, NJ. All rights reserved. This material is protected under all copyright laws as they currently exist. No portion of this material may be reproduced, in any form or by any means, without permission in writing from the publisher.

conductor. Thus, according to Gauss's law, there is no net charge inside, meaning that any net charge must lie on the surface.

Hints and Suggestions for Solving Problems

1. In most situations, charges on an object are expressed in microcoulombs (μC, 10^{-6} C). When using Coulomb's law for problems, such as in Section 15.3, make sure that the charges are in *Coulombs* and distance is in *meter*s. Convert the units, if necessary.

2. Coulomb's law is similar to Newton's law of gravitation. *Be mindful of the direction of the electric force* as given by Coulomb's law. The electric force can be *attractive* or *repulsive*. Draw the sketch showing the direction of the force. Take right and upward as positive directions and stick to this convention throughout the problem.

3. For more than two charges present in a problem, the net electric force is the *vector sum* of the forces between any pair. Use a convenient coordinate system, and sketch the forces. Use the method of components when adding two or more forces. Recall that the components of the resulting force vector are the sum of the components of the individual force vectors.

© 2007 Pearson Education, Inc., Upper Saddle River, NJ. All rights reserved. This material is protected under all copyright laws as they currently exist. No portion of this material may be reproduced, in any form or by any means, without permission in writing from the publisher.

4. Use the definition, $\vec{E} = \dfrac{\vec{F}_{\text{on } q+}}{q_+}$, to calculate electric field if the force on a test charge is known. Remember that the test charge is always of the positive kind. The preceding definition can be used also to find the force on any charge (positive or negative) if the electric field is known.

5. The electric field due to a point charge can be calculated from the expression $E = kq/r^2$. Watch the direction of the field. Use the method of components to determine the total electric field due to more than one charge, such as in the problems in Section 15.4.

6. When you draw a sketch of an electric field, remember that electric field lines are directed away from a positive charge and toward a negative charge. The force on a *positive charge* is in the *direction of the field,* whereas the force on a *negative charge* is *opposite* to the field.

7. Gauss's law is very useful for determining the electric field in situations having simple symmetry. Remember that if the net charge enclosed is zero, the electric field is also zero.

8. To determine an electric field using Gauss's law, such as in the problems in Section 15.6, determine the total number of electric field lines passing through a suitably chosen surface. Choose the symmetrical surface so that the electric field is normal to the surface. Calculate the total charge enclosed by the surface, and then apply Gauss's law.

©2007 Pearson Education, Inc., Upper Saddle River, NJ. All rights reserved. This material is protected under all copyright laws as they currently exist. No portion of this material may be reproduced, in any form or by any means, without permission in writing from the publisher.

CHAPTER 16
ELECTRIC POTENTIAL, ENERGY, AND CAPACITANCE

This chapter introduces the concept of electric potential energy, potential difference, and capacitance. It describes the laws of equivalent capacitance for series and parallel wiring of capacitors.

Key Terms and Concepts

> Electric potential energy
> Electric potential difference
> Volt
> Electric potential due to point charges
> Equipotential surface
> Capacitance
> Farad
> Parallel plate capacitor
> Dielectrics
> Equivalent series and parallel capacitances

16.1 Electric Potential Energy and Electric Potential Difference

It is useful to extend the concepts of mechanics such as work, kinetic energy, potential energy, and the work–energy theorem to electric cases to study electric fields.

Electric Potential Energy
Like the force of gravity, the electric force is a conservative force. As a result, there is an **electric potential energy**, U_e, associated with the electric force.

184

© 2007 Pearson Education, Inc., Upper Saddle River, NJ. All rights reserved. This material is protected under all copyright laws as they currently exist. No portion of this material may be reproduced, in any form or by any means, without permission in writing from the publisher.

The change in electric potential energy, ΔU_e, is given by $\Delta U_e = -W$, where W is the work done by the electric field. For example, if a positive charge q_+ is moved from plate A to plate B (separated by a distance d from plate A) of a parallel plate capacitor whose inside uniform electric field is E, the change in electric potential energy is

$$\Delta U_e = U_B - U_A = q_+Ed$$

This is analogous to the change in gravitational potential energy of a mass m when it is raised to a vertical distance h, as given by

$$\Delta U_g = U_B - U_A = mgh$$

Note that as in the gravitational case, the value of the potential energy at A or B or any point between is arbitrary. It is customary to take the potential energy at the negative plate of the capacitor to be zero.

Electric Potential Difference

The **electric potential difference** between two points in space is given by the change in electric potential energy *per unit positive test charge*. Thus, the change in electric potential, ΔV, is

$$\Delta V = \frac{\Delta U_e}{q_+} = \frac{W}{q_+} \tag{16.1}$$

The SI unit of electric potential difference is joule/coulomb (J/C) or volt (V).

Clearly, potential difference does not depend on the test charge that is moved. The potential difference for a

185

©2007 Pearson Education, Inc., Upper Saddle River, NJ. All rights reserved. This material is protected under all copyright laws as they currently exist. No portion of this material may be reproduced, in any form or by any means, without permission in writing from the publisher.

uniform electric field E between the two plates of a parallel plate capacitor is

$$\Delta V = \frac{\Delta U_e}{q_+} = \frac{q_+ Ed}{q_+} = Ed \qquad (16.2)$$

Thus, ΔV depends only on E and d. The positively charged plate is said to be at a higher electric potential than the negatively charged plate by ΔV. Thus, electric potential, V, at a point is the electric potential energy of a unit charge.

Note that electric potential difference is a more physically meaningful quantity than electric potential since the electric potential difference can be measured. The choice of electric potential as zero at the negative plate of a parallel plate capacitor or in some situations at infinity is arbitrary. However, in any situation the potential difference between two points is independent of the choice of zero for electric potential or the path followed by the test charge in moving from one point to the other point.

If an electron (carrying negative charge) is allowed to move freely in a parallel plate capacitor, it will fall toward the positive plate, the region of higher potential. Thus, *positive charges, when released, tend to move toward regions of low potential, and negative charges tend to move toward regions of high potential.*

Electric Potential Difference Due to a Point Charge
The potential difference (voltage) between any two points A and B due to a point charge q is

© 2007 Pearson Education, Inc., Upper Saddle River, NJ. All rights reserved. This material is protected under all copyright laws as they currently exist.
No portion of this material may be reproduced, in any form or by any means, without permission in writing from the publisher.

$$\Delta V = \frac{kq}{r_B} - \frac{kq}{r_A} \qquad (16.3)$$

where r_A and r_B are the distance of the points A and B, respectively, from the charge. Here, point B is closer to the charge and hence is at a higher potential. Keep in mind the following rules:

Electric potential increases $(+\Delta V)$ as we consider locations nearer to positive charges or farther from negative charges.

Electric potential decreases $(-\Delta V)$ as we consider locations farther from positive charges or nearer to negative charges.

If we choose the electric potential to be zero at infinity, the electric potential at a distance r from a point charge q is

$$V = \frac{kq}{r} \qquad (16.4)$$

Electric Potential Energy of Various Charge Configurations

The electric potential energy of two electric charges q_1 and q_2 separated by a distance r_{12} is

$$U_{12} = \frac{kq_1 q_2}{r_{12}} \qquad (16.5)$$

For *unlike* charges, since the electric force is attractive, the electrostatic potential energy is negative (such as gravitational potential energy). For *like* charges, on the other hand, potential energy is positive.

©07 Pearson Education, Inc., Upper Saddle River, NJ. All rights reserved. This material is protected under all copyright laws as they currently exist. No portion of this material may be reproduced, in any form or by any means, without permission in writing from the publisher.

Electric potential energy is a scalar quantity. The total electric potential energy due to several point charges is equal to the sum of the potential energies due to each pair of the charges. Thus,

$$U = U_{12} + U_{23} + U_{13} + \dots \qquad (16.6)$$

16.2 Equipotential Surfaces and the Electric Field

Equipotential Surfaces.
An **equipotential surface** is a surface on which electric potential is constant at any point. Thus, for the points A and A' on an equipotential surface,

$$V_A = V_{A'} \qquad (16.7)$$

Different equipotential surfaces have different electric potentials. For a parallel-plate capacitor each plate and any plane parallel to the plates are equipotential surfaces. Since any two points on an equipotential surface has the same potential, no work is done to move a charge from one point to the other on the equipotential surface. Since no work is done to move a charge along an equipotential surface, equipotential surfaces and the electric field lines are perpendicular to one another. Also, since electric field is a conservative field (like gravitational field), work done in moving a charge from one equipotential surface to the other is independent of the path followed by the charge. Equipotential surfaces cannot intersect one another. Electric field depends on the rate of change of the electric potential. If the electric potential changes by an amount ΔV for a displacement Δx in the direction of the maximum field change, the magnitude of the electric field is

© 2007 Pearson Education, Inc., Upper Saddle River, NJ. All rights reserved. This material is protected under all copyright laws as they currently exist. No portion of this material may be reproduced, in any form or by any means, without permission in writing from the publisher.

$$E = \left|\frac{\Delta V}{\Delta x}\right|_{max} \qquad (16.8)$$

The direction of the electric field is the direction in which the potential decreases most rapidly. If the electric field E is constant (such as between the plates of a parallel-plate capacitor), the potential V changes linearly with distance.

The SI unit of E is V/m or N/C. 1 N/C = 1 V/m

Since charges are free to move in an ideal conductor, no work is done. As a result, every point on or within the conductor is at the same potential. Ideal conductors are equipotential surfaces. The electric field, therefore, is directed perpendicularly to the surface of an ideal conductor.

The Electron Volt
An electron volt (eV) is the unit of energy usually used to express energies on the atomic scale. An **electron volt** is defined as the energy acquired by an electron or proton accelerated through a potential difference of 1 V.

$$1 \text{ eV} = 1.60 \times 10^{-19} \text{ J}.$$

In nuclear physics and particle physics, energies are also expressed in *kilo*electron volts (keV, 10^3 eV), *mega*electron volts (MeV, 10^6 eV), and *giga*electron volts (GeV, 10^9 eV).

16.3 Capacitance

A **capacitor** is a device that has a certain capacity to store electric charges. A simple capacitor consists of two parallel plates separated by a finite distance. The work

189

2007 Pearson Education, Inc., Upper Saddle River, NJ. All rights reserved. This material is protected under all copyright laws as they currently exist. No portion of this material may be reproduced, in any form or by any means, without permission in writing from the publisher.

needed to charge the plates could be done quickly by a battery. The battery removes electrons from the positive plate, pumps electrons to the negative plate, and does work in this process. The work done by the battery is stored as the potential energy in the electric field of the plates. The amount of charge, Q, on each plate is proportional to the potential difference, V, between the plates. Thus,

$$Q = CV \qquad \text{or} \qquad C = \frac{Q}{V} \qquad (16.9)$$

Here the constant C is called the **capacitance**, which is the charge stored per volt.

The SI unit of capacitance is coulomb/volt, which is called the farad (F).

$$1F = 1 \text{ C/V}$$

A farad is a large unit. Usually, capacitors with a capacitance of microfarad ($1 \mu F = 10^{-6}$ F) or picofarad (1 pF $= 10^{-12}$ F) are used in electric circuits.

Capacitance depends only on the geometry of the plates. For a parallel plate capacitor with an area of each surface, A, and separation between the plates, d, its capacitance is

$$C = \left(\frac{1}{4\pi k} \right) \frac{A}{d} \qquad (16.10)$$

The constant $\varepsilon_0 = \dfrac{1}{4\pi k} = 8.85 \times 10^{-12}$ C²/(N·m²)   (16.11)

is called the *permittivity of free space*. It describes the electrical properties of free space. Thus,

$$C = \frac{\varepsilon_0 A}{d} \qquad (16.12)$$

190

© 2007 Pearson Education, Inc., Upper Saddle River, NJ. All rights reserved. This material is protected under all copyright laws as they currently exist. No portion of this material may be reproduced, in any form or by any means, without permission in writing from the publisher.

A capacitor stores not only electric charges, it also stores electrical energy. Since $Q = CV$, a plot of voltage versus charge for a charging capacitor is a straight line with slope $1/C$ as shown. The average voltage is $V/2$.

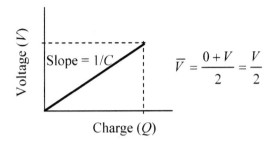

$$\overline{V} = \frac{0 + V}{2} = \frac{V}{2}$$

Thus, the total work done in charging the capacitor or the electric energy stored in the capacitor is

$$U_C = W = \frac{1}{2} QV$$

The expression can be written as

$$U_C = \frac{1}{2} QV = = \frac{1}{2} \frac{Q^2}{C} = \frac{1}{2} CV^2 \qquad (16.13)$$

16.4 Dielectrics

In most capacitors an electrically insulating material, called a **dielectric**, is inserted between the plates of a capacitor. It increases the charge storing capacity of the capacitor and, hence, the energy stored in the capacitor. This capability depends on the material and is expressed by the **dielectric constant** (κ) of the material.

2007 Pearson Education, Inc., Upper Saddle River, NJ. All rights reserved. This material is protected under all copyright laws as they currently exist. No portion of this material may be reproduced, in any form or by any means, without permission in writing from the publisher.

The applied field polarizes the molecules of the dielectric. This results in the alignment of positive charges on the surface of the dielectric near the negative plate, and negative charges on the surface near the positive plate. As a result, fewer electric field lines exist within the dielectric. The dielectric constant κ is given by

$$\kappa = \frac{V_0}{V} = \frac{E_0}{E} \qquad (16.14)$$

where E_0 and E are the electric fields and V_0 and V are the voltages across the plates before and after the dielectric has been added, respectively.

κ is dimensionless and is always greater than 1. In the presence of a dielectric

$$C = \kappa C_0 \qquad (16.15)$$

For a parallel plate capacitor with a dielectric,

$$C = \kappa C_0 = = \frac{\kappa \varepsilon_0 A}{d} \qquad (16.16)$$

© 2007 Pearson Education, Inc., Upper Saddle River, NJ. All rights reserved. This material is protected under all copyright laws as they currently exist. No portion of this material may be reproduced, in any form or by any means, without permission in writing from the publisher.

16.5 Capacitors in Series and in Parallel

Capacitors can be connected *in series* or *in parallel*.

Capacitors in Series

When capacitors are connected head to tail, one after another, so that there is the same charge on each, they are said to be in series.

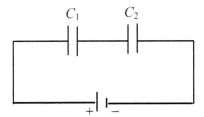

In this case, the voltage drops across all the capacitors is equal to the voltage of the source battery (V).

$$V = V_1 + V_2 + \ldots$$

Since $V_1 = Q/C_1$, $V_2 = Q/C_2, \ldots$ and $V = Q/C_s$, it can be shown that the **equivalent series capacitance**, C_s, of capacitors C_1, C_2, C_3, $\ldots$, connected in series is

$$\frac{1}{C_s} = \frac{1}{C_1} + \frac{1}{C_2} + \frac{1}{C_3} + \ldots \qquad (16.17)$$

Capacitors in Parallel

When capacitors are connected across the same voltage, they are said to be in parallel.

193

©2007 Pearson Education, Inc., Upper Saddle River, NJ. All rights reserved. This material is protected under all copyright laws as they currently exist. No portion of this material may be reproduced, in any form or by any means, without permission in writing from the publisher.

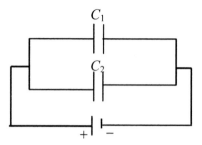

In this case, the total voltage

$$V = V_1 = V_2 + \ldots$$

The total charge Q_{total} is given by

$$Q_{\text{total}} = Q_1 + Q_2 + \ldots$$

Since $Q_1 = C_1V$, $Q_1 = C_1V$, etc. and $Q_{\text{total}} = C_pV$, the **equivalent parallel capacitance**, C_p, of capacitors C_1, C_2, C_3, … connected in parallel is

$$C_p = C_1 + C_2 + C_3 + \ldots \tag{16.18}$$

Hints and Suggestions for Solving Problems

1. The letter V is used to represent both the electric potential and its unit of measurement. Also the letter C is used for the unit of charge (coulomb) and for capacitance. Make sure the notations are not confusing.

2. When calculating potential energy of a charge, be mindful of its sign. An electron (negative charge) at a

194

© 2007 Pearson Education, Inc., Upper Saddle River, NJ. All rights reserved. This material is protected under all copyright laws as they currently exist. No portion of this material may be reproduced, in any form or by any means, without permission in writing from the publisher.

positive potential has a negative potential energy. A positive charge falls to a lower potential, whereas a negative charge falls to a higher potential.

3. Note that the unit of electric field can be expressed in N/C or V/m. The units are equivalent.

4. Electric potential and electric potential energy are scalars. As a result, when calculating the total electric potential or total electric potential energy due to several point charges algebraically, add the individual potentials or the potential energies, using the *appropriate signs*.

5. The directions of electric field and equipotential surfaces are always perpendicular. Use this information when drawing equipotential surfaces or when finding the direction of the electric field.

6. An electron volt (eV) is equivalent to 1.60×10^{-19} J. Be sure to convert the unit to joules (J), if the energy is given in electron volts.

7. The capacitance, C, is always a positive quantity. Always use the magnitudes of Q and V to determine C from the expression $C = Q/V$.

8. In numerical problems capacitance is usually given in microfarads or picofarads. Be sure to convert them into farads before solving any problem, such as those in Sections 16.3 and 16.4.

9. A dielectric always increases the capacitance of a capacitor. Use this information when solving problems in Section 16.4.

195

2007 Pearson Education, Inc., Upper Saddle River, NJ. All rights reserved. This material is protected under all copyright laws as they currently exist. No portion of this material may be reproduced, in any form or by any means, without permission in writing from the publisher.

CHAPTER 17
ELECTRIC CURRENT AND RESISTANCE

This chapter introduces the basic concepts of electricity including direct current, resistance, resistivity, Ohm's law, and electric power.

Key Terms and Concepts

Electric current
Direct current
Conventional current
Ampere
Drift velocity
Resistance
Ohm's law
Ohm
Resistivity
Temperature coefficient of resistivity
Electric power
Joule heat
Kilowatt-hours

17.1 Batteries and Direct Current

Water can flow through a hose when there is a pressure *difference*, and heat can flow from one object to another when there is a temperature *difference*. Similarly, electric charges can flow through a conductor when there exists a potential *difference* (commonly called voltage) across the ends of the conductor. The flow of electric charges produces an *electric current*. In a solid conductor positively charged nuclei are tightly bound, whereas the

196

© 2007 Pearson Education, Inc., Upper Saddle River, NJ. All rights reserved. This material is protected under all copyright laws as they currently exist. No portion of this material may be reproduced, in any form or by any means, without permission in writing from the publisher.

outer electrons are free to move when there is a potential difference. *Power supply* is the general name of a device that can produce potential difference.

Battery Action

A **battery** is a device that converts stored *chemical* energy into electric energy. A simple battery consists of two *electrodes* in an *electrolyte* solution that conducts electricity. With the suitable combination of electrodes and an electrolyte a potential difference develops between the electrodes because of a chemical reaction. In solid conductors, outer electrons of its atoms are relatively free to move. The potential difference between the electrodes causes a flow of charges (that is, causes an electric current) in a solid conductor (such as a metal wire) when connected to a battery. Current also flows in the electrolyte as positive ions migrate, completing the circuit.

The electrode that has the large number of excess electrons is called the **cathode** and designated as the negative (−) terminal of the battery, whereas the other electrode is called the **anode** and designated as the positive (+) terminal of the battery. The anode is at a higher potential than the cathode. As long as the internal chemical reaction maintains a potential difference cross the terminals of the battery, the electrons in the wire are pushed to flow in one direction in the circuit.

Battery EMF and Terminal Voltage

The potential difference across the terminals of a battery when it is not connected to a circuit is called the **electromotive force (emf)** of the battery. When a battery is connected to a circuit and current flows, the potential difference across the battery terminals is slightly less than the emf of the battery because of the *internal resistance* of

197

2007 Pearson Education, Inc., Upper Saddle River, NJ. All rights reserved. This material is protected under all copyright laws as they currently exist. No portion of this material may be reproduced, in any form or by any means, without permission in writing from the publisher.

the battery. The operating voltage of a battery is called the **terminal voltage**. The internal resistance of most batteries is typically small, thus the terminal voltage is nearly the same as the emf.

When batteries are connected in *series* (that is, the negative terminal of one battery is connected to the positive terminal of the next), the voltages are added. For example, a 12-volt automobile battery consists of six 2-volt cells connected in series. When cells are connected in *parallel* (that is, all positive terminals are connected to one point and all negative terminals are connected to another point), the potential difference is the same across all of them, and each cell supplies a fraction of the total current in the circuit.

Circuit Diagrams and Symbols
In analyzing electric circuits, different elements (such as a battery, etc.) are usually represented by their own symbols. Some common symbols used in electric circuits are:

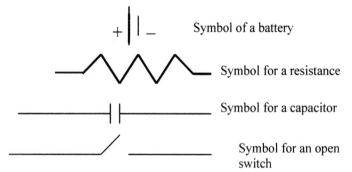

+ | | −   Symbol of a battery

Symbol for a resistance

Symbol for a capacitor

Symbol for an open switch

17.2 Current and Drift Velocity

A battery (or any source of voltage) connected to a continuous conducting path constitutes a **complete circuit**.

198

© 2007 Pearson Education, Inc., Upper Saddle River, NJ. All rights reserved. This material is protected under all copyright laws as they currently exist. No portion of this material may be reproduced, in any form or by any means, without permission in writing from the publisher.

A circuit usually has a switch to close or open the circuit path.

Electric Current

When a conductor is connected to a battery, electrons flow in the conductor toward the positive terminal of the battery. The direction of **conventional current** in a circuit is the direction in which a positive charge would move in the circuit. Clearly, positive charges move in the circuit from the positive terminal of the battery (region of high potential) to the negative terminal (region of low potential). As long as the internal chemical reaction maintains a potential difference across the terminals of the battery, the electrons in the wire are pushed to flow in one direction in the circuit. This type of current, where current does not change its direction, is called **direct current (dc)**.

Quantitatively, electric current is defined as the time rate of flow of net charges. If a net charge of amount q passes a given point in a time interval t, the electric current, I, is defined as

$$I = \frac{q}{t} \tag{17.1}$$

The SI unit of current is coulomb/second (C/s), or **ampere** (**A**). Small currents are expressed in *milliamperes* (mA, or 10^{-3} A), or *microamperes* (μA, or 10^{-6} A).

Drift Velocity, Electron Flow, and Electric Energy Transmission

In a solid conductor, the outermost electrons move in random directions and collide with the atoms of the lattice many times in one second. As a result, there is no net flow of charges in any direction, and the current is zero. When a potential difference is applied across the ends of the

199

2007 Pearson Education, Inc., Upper Saddle River, NJ. All rights reserved. This material is protected under all copyright laws as they currently exist. No portion of this material may be reproduced, in any form or by any means, without permission in writing from the publisher.

conductor, electrons still collide with the atoms as they move in different directions. But in this situation there is an added component of their velocities in a direction opposite to the electric field (that is, toward the high potential terminal of the battery). This net flow is characterized by an average velocity, called the **drift velocity,** of the electrons.

The magnitude of the drift velocity is about 1 mm/s, which is small. However, when a potential difference is applied, the associated electric field in the conductor moves at a speed of about 10^8 m/s (which is comparable to the speed of light). This electric field influences the motion of electrons almost instantaneously *throughout the conductor*, and, thus, a current is established in the circuit almost instantaneously.

17.3 Resistance and Ohm's Law

An applied voltage between the ends of a conducting material causes electrons to move. As electrons move through a conductor they suffer collisions with the atoms. These collisions create a **resistance** to the motion of electrons. In other words, because of its resistance, a substance opposes the flow of charges through it.

Resistance is given by the ratio of the voltage to the resulting current. That is, if the applied voltage is V to generate a current, I, in a conductor of resistance, R,

$$R = \frac{V}{I} \tag{17.2}$$

R is approximately constant for many materials over a range of voltages. For such materials the plot of V versus I

200

© 2007 Pearson Education, Inc., Upper Saddle River, NJ. All rights reserved. This material is protected under all copyright laws as they currently exist. No portion of this material may be reproduced, in any form or by any means, without permission in writing from the publisher.

is a straight line. The preceding relation when R is constant is known as **Ohm's law**.

The unit of resistance is the **ohm** (Ω). $1\ \Omega = 1\text{V/A}$.

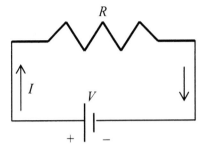

Ohm's law is not a fundamental law. Semiconductors show a *nonlinear* relationship between voltage and current and do not obey Ohm's law.

Factors That Influence Resistance
Since resistance arises because of collisions with lattice atoms, it is a material property and depends on the geometric factors of the material. The resistance of a material is *directly* proportional to the length (L) and *inversely* proportional to its cross-sectional area (A):

$$R \propto \frac{L}{A}$$

Resistivity
The resistance of a given material can be written as

$$R = \rho\,\frac{L}{A} \qquad (17.3)$$

201

2007 Pearson Education, Inc., Upper Saddle River, NJ. All rights reserved. This material is protected under all copyright laws as they currently exist. No portion of this material may be reproduced, in any form or by any means, without permission in writing from the publisher.

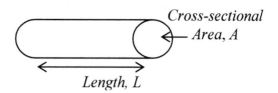

Cross-sectional Area, A

Length, L

The quantity ρ is a characteristic of the material called **resistivity**. The SI unit of resistivity is ohm·meter (Ω·m). From Equation 17.3,

$$\rho = \frac{RA}{L}$$

The resistivity of a conductor increases almost linearly with its temperature. That is, the resistivity ρ at temperature T is given by:

$$\rho = \rho_0 (1 + \alpha \Delta T) \qquad (17.4)$$

where ρ_0 is the resistivity at temperature T_0 and $\Delta T = T - T_0$. The constant α is called the temperature coefficient of resistivity. Clearly, the change in resistivity is

$$\Delta \rho = \rho_0 \alpha \Delta T \qquad (17.5)$$

Also,

$$R = R_0 (1 + \alpha \Delta T) \quad \text{or} \quad \Delta R = R_0 \alpha \Delta T \quad (17.6)$$

Semiconductors have negative α; that is, their resistances decrease with increasing temperature.

Superconductivity

There is an important class of materials whose resistivity suddenly drops to zero below a certain temperature, called the critical temperature, T_C. This is called

202

© 2007 Pearson Education, Inc., Upper Saddle River, NJ. All rights reserved. This material is protected under all copyright laws as they currently exist. No portion of this material may be reproduced, in any form or by any means, without permission in writing from the publisher.

superconductivity. Below its critical temperature the materials that show superconductivity are called **superconductors** (such as mercury below 4.2 K). Superconductivity is a result of quantum effects. The highest temperature at which superconductivity has been observed recently is about 125 K. "High-T_c" superconductors are presently of research interest because of their potential applications.

17.4 Electric Power

Energy must be supplied for an electric current to flow through a circuit. The average rate at which electrical energy is supplied is called average **electric power**. If the applied potential difference is V, the average electric power, $\overline{P}$, spent is

$$\overline{P} = \frac{W}{t} = \frac{qV}{t}$$

where W is the work done by the voltage source to move the charge q. Average power is the same as the power at all times when current and voltage do not change with time. Thus, the preceding power expression becomes, in terms of current I,

$$P = IV \qquad\qquad (17.7a)$$

The SI unit of electric power is the watt (W).

Electric power is dissipated in the form of heat in a resistor. As the charges pass through the conductor they collide with the lattice atoms. Energy is transferred by the moving charge in collisions with the atoms, resulting in a rise of temperature of the conductor.

203

2007 Pearson Education, Inc., Upper Saddle River, NJ. All rights reserved. This material is protected under all copyright laws as they currently exist. No portion of this material may be reproduced, in any form or by any means, without permission in writing from the publisher.

If the resistor R is connected across a potential difference, V, the power dissipated is (since $V = RI$)

$$P = IV = \frac{V^2}{R} = I^2R \qquad (17.7b)$$

Joule Heat

The thermal energy dissipated in a current-carrying resistor is called **Joule heat** or I^2R **losses**. Joule heating is undesirable in some situations, such as in electrical transmission lines. In other situations, such as in heaters and toasters, conversion of electrical energy to thermal energy is important. In electric light bulbs the electric energy is consumed to radiate as light and heat.

A practical unit of electric energy is the **kilowatt-hour (kWh)**. 1 kilowatt-hour = 1000 W X 3600 s = 3.6×10^6 J Electric companies determine electric bills based on the number of kilowatt-hours of energy used.

Electrical Efficiency and Natural Resources

The need for the use of electricity is continuously increasing. In the United States about 25% of the generated electricity goes into lighting. Because the consumption of electricity is huge, the federal and many state governments set a minimum efficiency limit for domestic electronics and appliances. Researchers are using new techniques to develop more efficient appliances and lighting. Saving electric energy directly means saving fuels and natural resources and a reduction in natural hazards such as global warming.

© 2007 Pearson Education, Inc., Upper Saddle River, NJ. All rights reserved. This material is protected under all copyright laws as they currently exist. No portion of this material may be reproduced, in any form or by any means, without permission in writing from the publisher.

Hints and Suggestions for Solving Problems

1. Remember that the same letter, C, is used for unit of charge (coulomb) and for capacitance. Also, the letter V is used for volt and for electric potential.

2. The electric current, I, is treated as a positive quantity. When using the definition $I = q/t$ to determine current, use the magnitude of the charge q.

3. $R = \rho L/A$, that is, R is linearly proportional to the length L and inversely proportional to the area A.

4. Remember the resistivity of a conductor increases with its temperature as $\rho = \rho_0 (1 + \alpha \Delta T)$. Similarly, $R = R_0 (1 + \alpha \Delta T)$.

5. Use of Ohm's law $(I = V/R)$ and determination of power $(P = IV = I^2 R = V^2/R)$ dissipated through a resistor is straightforward. Use Ohm's law, the expression for resistivity, and the expressions for power to solve problems in Sections 17.3 and 17.4.

2007 Pearson Education, Inc., Upper Saddle River, NJ. All rights reserved. This material is protected under all copyright laws as they currently exist. No portion of this material may be reproduced, in any form or by any means, without permission in writing from the publisher.

CHAPTER 18
BASIC ELECTRIC CIRCUITS

This chapter discusses series, parallel, series-parallel combinations of resistances, and Kirchhoff's rules for circuit analysis. It also discusses how ammeters and voltmeters work, and discusses household circuits and electrical safety.

Key Terms and Concepts

Resistors in series and parallel
Equivalent resistance
Series-parallel resistor combination
Kirchhoff's rules
RC circuits
Time constant
Ammeter
Voltmeter
Galvanometer
Fuse
Circuit breaker
Grounded plugs
Polarized plug

18.1 Resistances in Series, Parallel, and Series-Parallel Combinations

An electric circuit may contain any number of resistors connected in different ways.

Resistors in Series
When resistors are connected one after another so that the same current flows through each of them, they are said to be in **series**.

206

© 2007 Pearson Education, Inc., Upper Saddle River, NJ. All rights reserved. This material is protected under all copyright laws as they currently exist. No portion of this material may be reproduced, in any form or by any means, without permission in writing from the publisher.

Series combination

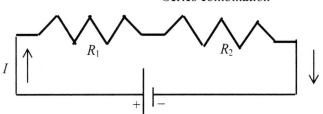

Keep in mind that the *sum of the voltages* (with proper signs) *around a circuit loop is zero.* For resistors connected in series with a battery of voltage V,

$$V - \Sigma \, IR_i = 0 \qquad (18.1)$$

Or, $V = IR_1 + IR_2 + IR_3 + ... = I(R_1 + R_2 + R_3 + ...) = IR_s$

Thus, the **equivalent series resistance**, R_s, is

$$R_s = R_1 + R_2 + R_3 + ... = \sum R_i \qquad (18.2)$$

Clearly, the equivalent resistance is greater than the individual resistances. In a series circuit, if one resistor is broken, the whole combination is open because current cannot flow.

Resistors in Parallel

When resistors are connected so that all the leads on one side of the resistors are attached together, as are the leads on the other side, they are said to be in **parallel**.

2007 Pearson Education, Inc., Upper Saddle River, NJ. All rights reserved. This material is protected under all copyright laws as they currently exist. No portion of this material may be reproduced, in any form or by any means, without permission in writing from the publisher.

Parallel combination

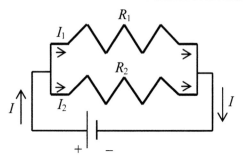

In this case, *the voltage drop across each resistor is the same*, and the current has parallel paths through which to flow. The total current is equal to the sum of the currents through the individual resistors:

$$I = I_1 + I_2 + I_3 + ...$$

It can be shown that the **equivalent parallel resistance**, R_p, for resistors connected in parallel is given by

$$\frac{1}{R_p} = \frac{1}{R_1} + \frac{1}{R_2} + \frac{1}{R_3} + ... = \sum \left(\frac{1}{R_i} \right) \qquad (18.3)$$

If there are only two resistors in parallel,

$$R_p = \frac{R_1 R_2}{R_1 + R_2} \qquad (18.3a)$$

The equivalent resistance for resistors in parallel is *always less than the smallest resistance in the combination*. Resistance is less because the current has different paths through which to flow.

208

© 2007 Pearson Education, Inc., Upper Saddle River, NJ. All rights reserved. This material is protected under all copyright laws as they currently exist. No portion of this material may be reproduced, in any form or by any means, without permission in writing from the publisher.

The general rules of thumb are:
Series connection increases total resistance.
Parallel connection decreases total resistance.

Series–Parallel Resistor Combinations
Resistors are often wired partially in series and partially in parallel in a circuit. In such a complex circuit, the usual steps for analysis are as follows:

1. Starting with the resistor combination farthest from the voltage source, find the equivalent series and parallel resistances.

2. Reduce the circuit to a single loop with one equivalent resistance.

3. Find the total current using $I = V/R_{total}$.

4. Expand the circuit by reversing the reduction steps, and use the current for the reduced circuit to find the currents and voltages for the resistors in each step.

18.2 Multiloop Circuits and Kirchhoff's Rules

The following two **Kirchhoff rules** are very useful for analyzing multiloop electric circuits. These rules basically express conservation of charge (the junction theorem) and conservation of energy (the loop theorem) in a closed electric circuit.

Kirchhoff's Junction Theorem
It states that the algebraic sum of all currents meeting at a junction is zero. That is,

2007 Pearson Education, Inc., Upper Saddle River, NJ. All rights reserved. This material is protected under all copyright laws as they currently exist. No portion of this material may be reproduced, in any form or by any means, without permission in writing from the publisher.

$$\sum I_i = 0. \qquad (18.4)$$

The current entering a junction is considered to be positive, and current leaving a junction is considered to be negative.

Kirchhoff's Loop Theorem
The algebraic sum of all potential differences (voltages) across each element around any *closed* circuit loop is zero. That is,

$$\sum V_i = 0. \qquad (18.5)$$

Voltage *increases* in moving from the negative to the positive terminal of a battery and *decreases* in crossing a resistor in the direction of the current.

Application of Kirchhoff's Rules
Use the following steps in applying Kirchhoff's rules.

1. Assign a current and current direction for each branch in the circuit.

2. Indicate the loops and the chosen directions to traverse the loops.

3. Apply Kirchhoff's loop theorem (first rule) for the equations for the currents, one for each junction.

4. Traverse the number of loops necessary to include all branches and apply Kirchhoff's loop theorem (second rule) with proper sign convention.

5. Solve the resulting equations for unknowns.

© 2007 Pearson Education, Inc., Upper Saddle River, NJ. All rights reserved. This material is protected under all copyright laws as they currently exist. No portion of this material may be reproduced, in any form or by any means, without permission in writing from the publisher.

18.3 *RC* Circuits

Circuits containing both resistors and capacitors are called
RC **circuits**.

Charging a Capacitor through a Resistor
For a circuit containing a resistor, R, and a capacitor, C,
both connected in series with a battery of emf V_0, the
voltage across the capacitor changes with time as

$$V_C = V_0 \left(1 - e^{-t/RC}\right) \qquad (18.6)$$

Note that when the switch is closed, immediately there is
no charge on the capacitor and thus no voltage across it, as
seen from Equation (18.6). With time, charge increases on
the plate and voltage increases until, at very large times,
the voltage across the capacitor is V_0 and the voltage
across the resistor is zero.

This relation shows that current is high initially (that is, the
capacitor behaves as a short circuit), then decreases to zero
as $t \to \infty$ (that is, in this limit, the capacitor behaves as an
open switch).

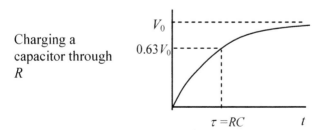

Charging a
capacitor through
R

The current changes with time as:

©2007 Pearson Education, Inc., Upper Saddle River, NJ. All rights reserved. This material is protected under all copyright laws as they currently exist.
No portion of this material may be reproduced, in any form or by any means, without permission in writing from the publisher.

$$I = I_0 e^{-t/RC} \qquad (18.7)$$

where I_0 is the initial current given by $I_0 = V/R$.

The product RC has the dimension of time; it is the characteristic time, called the **time constant** (τ) of the RC circuit:

$$\tau = RC \qquad (18.8)$$

As seen from Equation (18.6), at $t = \tau = RC$, the voltage across the capacitor has risen to $V_C \approx 0.63\ V_0$.

Discharging a Capacitor through a Resistor
When a fully charged capacitor is discharged through a resistor, the voltage across the capacitor decreases exponentially with time:

$$V_C = V_0 e^{-t/RC} = V_0 e^{-t/\tau} \qquad (18.9)$$

The current in the circuit also decreases exponentially according to Equation (18.7) as shown.

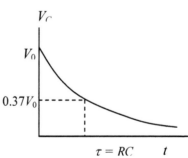

Discharging a capacitor through R

212

© 2007 Pearson Education, Inc., Upper Saddle River, NJ. All rights reserved. This material is protected under all copyright laws as they currently exist. No portion of this material may be reproduced, in any form or by any means, without permission in writing from the publisher.

18.4 Ammeters and Voltmeters

Ammeters and **voltmeters** are devices used to determine current and voltage (potential difference), respectively, in an electric circuit. The basic design of both of these is a **galvanometer**, which is a current-sensitive device whose needle deflection is proportional to the current through its coil.

The Ammeter

To measure current in a section of a circuit, an ammeter is connected in series with the section. The resistance of an ammeter is low compared with the resistances of the circuit; otherwise, it will alter the current it is intended to measure. To measure large currents (larger than microamperes), an ammeter must have a small *shunt resistor* in parallel with the coil. Because of the shunt resistor, the majority of the current bypasses the galvanometer.

The Voltmeter

To measure the potential difference across a section, a voltmeter is connected *in parallel* with the section. The resistance of a voltmeter must be very high compared with the resistances of the circuit; otherwise, it will draw a significant amount of current and alter the potential difference it is intended to measure. A voltmeter capable of measuring voltages higher than the microvolt range is constructed by connecting a large *multiplier resistor* in series with a galvanometer.

A single device that can function as an ammeter, a voltmeter, or an ohmmeter, is called a *multimeter*. This function is accomplished by providing a choice of several shunt or multiplier resistors. Electrical digital multimeters

213

2007 Pearson Education, Inc., Upper Saddle River, NJ. All rights reserved. This material is protected under all copyright laws as they currently exist. No portion of this material may be reproduced, in any form or by any means, without permission in writing from the publisher.

analyze digital signals to calculate a voltage, current, and resistance.

18.5 Household Circuits and Electrical Safety

The elements in household circuits (such as lamps, appliances, etc.) are connected in parallel so that when one element malfunctions the other elements in the circuit continue to work. The power rating is 120 V for appliances in the United States. Power is supplied by a three-wire system. There is a potential difference of 240 V between the two hot wires, and each of them has a potential difference of 120 V between the ground. The third wire is grounded and is called the *ground*, or *neutral*, wire. The 120 V potential difference needed for household appliances is obtained by connecting them between the ground wire and either hot wire. Large appliances (such as central air conditioners) that need 240 V for operation are connected between the two hot wires.

There is a limit to the number of elements that can be on simultaneously in a circuit. Adding more elements in a parallel circuit decreases the equivalent resistance and, hence, increases the total current in the circuit. Thus, by adding too many current-carrying elements, the heat produced by the current can be high enough to melt the insulation of the wire and catch fire. Adding fuses and circuit breakers in the circuit prevents this overloading. **Fuse**s have metal strips inside that melt because of joule heat and break the circuit when the current exceeds the limit. **Circuit breakers** contain bimetallic strips. As the current exceeds the limit, the bimetallic strip bends enough

214

© 2007 Pearson Education, Inc., Upper Saddle River, NJ. All rights reserved. This material is protected under all copyright laws as they currently exist. No portion of this material may be reproduced, in any form or by any means, without permission in writing from the publisher.

to open the circuit mechanically. The strip cools quickly and can be reset.

Switches, fuses, and circuit breakers are wired on the hot side of the line. If the elements were wired in the grounded side, the line would be at a high potential even when the fuse is blown or the switch is open. Even if the fuse is on the hot side a potentially dangerous situation exists. If an internal wire comes in contact with the metal casing of a power tool, a person touching the casing, which is at high voltage, can get a shock. To prevent a shock, a third dedicated grounding wire is added to the circuit to ground the metal casing. On three-prong **grounded plugs**, the large round prong connects with the dedicated grounding wire. **Polarized plugs** are two-prong plugs that fit in the sockets only one way. Wall receptacles are wired in such a way that the small slit connects to the hot side and the large slit connects to the neutral side. Using polarized plugs, manufacturers of appliances can make sure that the switch is on the hot side of the line and the casing of the appliance is on the grounded side.

2007 Pearson Education, Inc., Upper Saddle River, NJ. All rights reserved. This material is protected under all copyright laws as they currently exist. No portion of this material may be reproduced, in any form or by any means, without permission in writing from the publisher.

Hints and Suggestions for Solving Problems

1. A typical mistake made in calculating the equivalent resistance of two resistors in parallel is to compute $1/R_1$ and $1/R_2$ and then to add and present the sum as R_p rather than as $1/R_p$. Note that the sum of the reciprocals is not equal to the reciprocal of the sum. That is, $1/R_1 + 1/R_2$ is not equal to $1/(R_1 + R_2)$.

2. For two resistors in parallel, use the following expression to find the equivalent resistance: $R_p = R_1R_2/(R_1 + R_2)$.

3. Recall that the equivalent resistance of a number of resistors is smaller than the smallest of the individual resistors. For a complex combination of resistors, reduce them to a single resistor by repeatedly applying the series and parallel laws for resistors. Keep these in mind when solving problems in Section 18.1.

4. Apply Kirchhoff's junction theorem and loop theorem to solve the problems in Section 18.2. The theorems may also be useful for solving some problems in Section 18.1.

5. When applying the junction theorem, keep in mind that a current entering a junction is considered to be positive, and a current leaving a junction is considered to be negative. Also, when applying the loop theorem, keep in mind that the potential *increases* in moving from the negative to the positive terminal of a battery and *decreases* in crossing a resistor in the direction of the current.

216

© 2007 Pearson Education, Inc., Upper Saddle River, NJ. All rights reserved. This material is protected under all copyright laws as they currently exist. No portion of this material may be reproduced, in any form or by any means, without permission in writing from the publisher.

6. The time constant $\tau\,(= RC)$ has the dimension of time. If C is in farads and R is in ohms, the unit of τ will be in seconds.

7. The voltage and current in RC circuit vary as

   $V_c = V_0\left(1 - e^{-t/\tau}\right)$ and $I = I_0\,e^{-t/\tau}$, respectively. For a discharging capacitor $V_c = V_0\,e^{-t/\tau}$. Use these expressions to solve the problems in Section 18.3.

8. To solve problems on ammeters and voltmeters, such as those in Section 18.4, remember that an ammeter is connected in series in a circuit and has a low resistance. A shunt is usually connected in parallel with an ammeter to measure high current. A voltmeter is connected in parallel, has high resistance, and ideally draws no current. Usually, a high resistance is added in series in a voltmeter.

217

©2007 Pearson Education, Inc., Upper Saddle River, NJ. All rights reserved. This material is protected under all copyright laws as they currently exist. No portion of this material may be reproduced, in any form or by any means, without permission in writing from the publisher.

CHAPTER 19
MAGNETISM

This chapter describes magnetism and the relationship between magnetism and electricity. In particular, it describes the basics of magnetism, force acting on a charge moving in a magnetic field, force and torque acting on a current-carrying coil in a magnetic field, electromagnetism, and its applications.

Key Terms and Concepts

        Law of poles
        Magnetic field and magnetic field lines
        Electromagnetism
        Magnetic force
        Tesla
        Right-hand force rule
        Magnetic permeability
        Right-hand source rule
        Ferromagnetic materials
        Magnetic domains
        Magnetic force on a current-carrying wire
        Right-hand rule for a current-carrying wire
        Torque on a current-carrying loop
        Magnetic moment
        The galvanometer
        dc motor
        Cathode ray tube
        Mass spectrometer
        The electronic balance
        Earth's magnetic field
        Magnetic declination

© 2007 Pearson Education, Inc., Upper Saddle River, NJ. All rights reserved. This material is protected under all copyright laws as they currently exist.
No portion of this material may be reproduced, in any form or by any means, without permission in writing from the publisher.

19.1 Magnets, Magnetic Poles, and Magnetic Field Direction

Permanent magnets, such as a bar magnet or a magnetic needle, when suspended freely, always point in the north–south direction. The end of the magnet pointing toward the north is called the *north-seeking* or *north pole*, and the end pointing toward the south is called the *south pole*. The **pole-force law** or **law of poles** describes the interactions between the poles of magnets. According to the law, *the like magnetic poles repel each other, whereas unlike magnetic poles attract each other.*

Magnets always have two poles. A magnetic monopole (a single north or south pole) has not been observed. If you break a bar magnet in half, each half becomes two shorter magnets, still with two poles. This behavior differs from that of electric charges, because two types of charges can exist separately.

Magnetic Field Direction

Just as an electric charge creates an electric field, a magnet creates a **magnetic field** ($\vec{B}$) that surrounds every magnet. Like an electric field, a magnetic field is a vector quantity. Also, like an electric field, a magnetic field can be represented by its magnetic field lines. *The direction of the magnetic field $\vec{B}$ at a point is the direction in which the north pole of a compass needle points when placed at that point.* The magnetic field lines leave the north pole of a magnet and enter at the south pole. The field lines continue within the body of the magnet. Unlike electric fields, magnetic fields always form closed loops. Also, with an electric field, the magnetic field lines are dense where the magnetic field is strong. Magnetic field lines can be

2007 Pearson Education, Inc., Upper Saddle River, NJ. All rights reserved. This material is protected under all copyright laws as they currently exist. No portion of this material may be reproduced, in any form or by any means, without permission in writing from the publisher.

visualized by placing a sheet of paper on top of a bar magnet and dropping iron filings onto the paper. The iron filings align with the field lines.

19.2 Magnetic Field Strength and Magnetic Force

A magnetic field exerts a force on moving electric charges. The study of interactions between electrically charged particles and magnetic fields is called **electromagnetism**. Experimentally, it is found that the magnetic force is proportional to the particle's charge and its speed. When the velocity ($\vec{v}$) is perpendicular to the magnetic field ($\vec{B}$), the magnetic force exerted on the charge is used to define the magnetic field as

$$B = \frac{F}{qv} \qquad (19.1)$$

Thus, magnetic field is given by the magnetic force per unit charge per unit speed.

The SI unit of B is the **tesla (T)**. 1 tesla = $1 \text{N}/(\text{C} \cdot \text{m/s})$ = N/A·m. The tesla is a large unit. Another commonly used unit of magnetic fields is the gauss (G). 1 G = 10^{-4} T.

From Equation (19.1), the force on a charged particle (for $\vec{v}$ perpendicular to $\vec{B}$) is

$$F = qvB \qquad (19.2)$$

© 2007 Pearson Education, Inc., Upper Saddle River, NJ. All rights reserved. This material is protected under all copyright laws as they currently exist. No portion of this material may be reproduced, in any form or by any means, without permission in writing from the publisher.

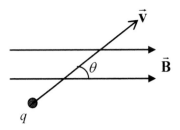

In general, the magnitude of the force, F, experienced by a charge, q, moving with a velocity $\vec{v}$ in a magnetic field $\vec{B}$ is given by

$$F = qvB \sin \theta \qquad (19.3)$$

where θ is the angle between $\vec{v}$ and $\vec{B}$.
Clearly, if $\theta = 0°$ or $180°$, $F = 0$. Also, if $v = 0$, $F = 0$.
F is a maximum when $\theta = 90°$.

The Right-Hand Force Rule for Moving Charges
The direction of the magnetic force on a positively charged particle can be obtained from the **right-hand force rule**:

If you point the fingers of your right hand in the direction of $\vec{v}$ and curl the fingers toward $\vec{B}$, your thumb will point in the direction of the force, $\vec{F}$, for a positive charge.

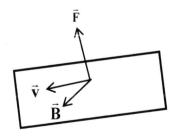

©2007 Pearson Education, Inc., Upper Saddle River, NJ. All rights reserved. This material is protected under all copyright laws as they currently exist.
No portion of this material may be reproduced, in any form or by any means, without permission in writing from the publisher.

If the moving charges are negative, the direction of the force is opposite to the direction that is given by the preceding rule.

19.3 Applications: Charged Particles in Magnetic Fields

A moving charge in a magnetic field experiences a force that depends on the particle's mass, charge, velocity, and the magnetic field. This force has applications in many common appliances.

The Cathode Ray Tube (CRT): Oscilloscope Screens, Television Sets, and Computer Monitors

The **cathode ray tube** (CRT) is a vacuum tube that is used in oscilloscopes, televisions, and computer monitors. The electrons are emitted from a hot filament in an electron gun and are accelerated by a high voltage between the cathode and anode and then deflected to a proper location on a fluorescent screen by a magnetic field produced by current-carrying coils. As the field strength is varied, the beam scans the screen in a fraction of a second. The fluorescent material of the screen emits light as the electron beam hits the material. In a black and white TV the image is reproduced on the screen as a mosaic of light and dark dots.

A color picture tube has three beams, one for each primary color—red, green, and blue. Phosphor dots on the screen are arranged in groups of three (one dot for each primary color). The excitation of the appropriate dots with the resulting combination of colors produces a color picture.

222

© 2007 Pearson Education, Inc., Upper Saddle River, NJ. All rights reserved. This material is protected under all copyright laws as they currently exist. No portion of this material may be reproduced, in any form or by any means, without permission in writing from the publisher.

The Velocity Selector and the Mass Spectrometer

In a **mass spectrometer** the *mass* of ions ise measured. Mass spectrometers are used to track short-lived molecules to study biochemistry of living organisms, to determine the structure of large organic molecules, and to analyze the composition of complex mixtures. They are also used to separate different isotopes of atoms, to determine the age of rocks and human artifacts by measuring the abundance of different isotopes, etc.

Heating the substance produces the ions. If the electric field, magnetic field, and the direction of motion of a charge are all mutually perpendicular in a region, it is possible that for a certain v (with proper direction) the electric force (qE) on the charge is equal and opposite to the magnetic force (qvB_1). This condition gives

$$v = \frac{E}{B_1}$$

If the plates are parallel, $E = V/d$, where V is the voltage between the plates and d is the distance between them. Thus,

$$v = \frac{V}{B_1 d} \tag{19.4}$$

A charge moving with this speed will not be deflected by the fields. This is the principle of a **velocity selector**.

In the spectrometer, the ions pass through a velocity selector with the known velocity v and enter a second magnetic field B_2. The ions are deflected in a circular path depending on the mass and net charge on the ion. It can be shown that the mass m is

223

2007 Pearson Education, Inc., Upper Saddle River, NJ. All rights reserved. This material is protected under all copyright laws as they currently exist. No portion of this material may be reproduced, in any form or by any means, without permission in writing from the publisher.

$$m = \left(\frac{qd\, B_1 B_2}{V} \right) r \qquad (19.5)$$

where $V = Ed$ (d is the separation between the plates in the electric field). Clearly, the greater the mass of the ion, the greater the radius r of the path. If the quantities on the right hand side of the preceding equation are known, the mass of an ion can be determined. The result, called the *mass spectrum* (where the number of ions versus their molecular mass is plotted), is displayed on an oscilloscope or computer screen.

Silent Propulsion: Magnetohydrodynamics
Magnetohydrodynamics is the study of the interaction of moving fluids and magnetic fields. The silent-running feature of the modern submarine is based on magnetohydrodynamics. In this case, a superconducting electromagnet produces a large magnetic field, and an electric generator that produces a large dc voltage sends current through the seawater. The magnetic force on the current pushes the water backward, while the submarine accelerates silently by reaction force in the forward direction.

19.4 Magnetic Forces on Current-Carrying Wires

The motion of charges causes an electric current. Since moving charges experience force in a magnetic field, a current-carrying wire also experiences a force in a magnetic field.

© 2007 Pearson Education, Inc., Upper Saddle River, NJ. All rights reserved. This material is protected under all copyright laws as they currently exist.
No portion of this material may be reproduced, in any form or by any means, without permission in writing from the publisher.

The magnitude of the force on a wire of length L carrying a current I in a direction perpendicular to the magnetic field, $\vec{B}$, is given by

$$F = ILB \qquad (19.6)$$

In general,

$$F = ILB \sin \theta \qquad (19.7)$$

where θ is the angle between the current and the magnetic field. F is maximum when $\theta = 90°$ and is zero when $\theta = 0°$ or $180°$.

The direction of the force is given by the **right-hand force rule for a current-carrying wire**.

When the fingers of the right hand are pointed in the direction of the current I and then curled toward the vector $\vec{B}$, the extended thumb points in the direction of the magnetic force on the wire.

Torque on a Current-Carrying Loop

A current-carrying loop experiences a torque when placed in a magnetic field. The amount of torque on a given loop depends on the orientation of the loop and the magnetic field. The torque, τ, exerted on a rectangular loop of area A carrying current I is

$$\tau = IAB \sin \theta \qquad (19.8)$$

where B is the magnetic field and θ is the angle between the magnetic field and the normal to the area of the loop.

The preceding relation for the torque for a rectangular loop is valid for a flat loop of any shape. If a coil has N number of single loops, the torque is

225

2007 Pearson Education, Inc., Upper Saddle River, NJ. All rights reserved. This material is protected under all copyright laws as they currently exist. No portion of this material may be reproduced, in any form or by any means, without permission in writing from the publisher.

$$\tau = NIAB \sin \theta \qquad (19.9)$$

The **magnetic moment** (m) of the loop is defined as

$$m = NIA \qquad (19.10)$$

and its unit is A·m^2. Thus,

$$\tau = mB \sin \theta \qquad (19.11)$$

The torque tends to align the magnetic moment vector **m** with the magnetic field direction. Torque is maximum when $\theta = 90°$ and is zero when $\theta = 0°$ or $180°$.

A spinning electron is equivalent to a concentric current loop and, thus, possesses a magnetic moment. When placed in a magnetic field this moment will tend to align with the field. In a ferromagnetic material, this produces a net magnetic field much stronger than the external field.

19.5 Applications: Current-Carrying Wires in Magnetic Fields

The Galvanometer: The Foundation of the Ammeter and Voltmeter

A galvanometer is an essential part of the design of an ammeter or voltmeter. A galvanometer consists of loops of wire on an iron core that pivots between the pole faces of a permanent magnet. The pole faces of the magnet are curved, and the coil is wrapped on a cylindrical iron core. When a current exists in the coil, the coil experiences a torque. A needle is attached to the coil. As the coil rotates, the magnetic field is always perpendicular to the

© 2007 Pearson Education, Inc., Upper Saddle River, NJ. All rights reserved. This material is protected under all copyright laws as they currently exist.
No portion of this material may be reproduced, in any form or by any means, without permission in writing from the publisher.

nonpivoted side of the coil. The deflection of the needle is proportional to the current.

The dc Motor

An electric motor converts electric energy to mechanical energy. A **dc motor** generates continuous rotation for continuous energy output. In a motor, the torque on the coil rotates the axle of the motor. As the rotor turns, a split-ring commutator reverses the direction of current to ensure that the torque is always in the same direction. For a real motor, the rotating shaft is called the *armature*.

The Electronic Balance

In a traditional balance, the mass of an object is determined by balancing the weight of the unknown mass to that of a known mass. In a digital balance there is a suspended beam with a pan at one end to hold the unknown mass. The balancing downward force is provided by the current-carrying coils placed inside a permanent magnet. The force is proportional to the current and the unknown mass is determined from the current that produces a force needed to balance the beam.

19.6 Electromagnetism: The Source of Magnetic Fields

A current-carrying conductor generates a magnetic field. This phenomenon was first discovered by Orsted in 1820. Orstd observed that as soon as a current was established in a wire, a nearby compass needle was deflected from its usual orientation. This key observation unified the theories of electricity and magnetism.

2007 Pearson Education, Inc., Upper Saddle River, NJ. All rights reserved. This material is protected under all copyright laws as they currently exist. No portion of this material may be reproduced, in any form or by any means, without permission in writing from the publisher.

Magnetic Field Near a Long, Straight, Current-Carrying Wire

At a perpendicular distance d from a long, straight wire carrying a current I, the magnitude of the magnetic field **B** is

$$B = \frac{\mu_0 I}{2\pi d} \tag{19.12}$$

where the constant $\mu_0 = 4\pi \times 10^{-7}$ T·m/A, called the *magnetic permeability of free space.*

The magnetic field lines due to a current in a long, straight wire are closed circles centered on the wire.

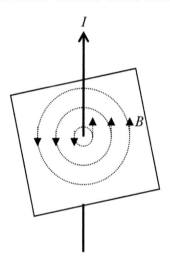

The direction of the magnetic field is given by the **right-hand source rule**.

Point the thumb of your right hand in the direction of the current, and curl the fingers into the shape of a half-circle.

228

© 2007 Pearson Education, Inc., Upper Saddle River, NJ. All rights reserved. This material is protected under all copyright laws as they currently exist.
No portion of this material may be reproduced, in any form or by any means, without permission in writing from the publisher.

The tips of the fingers will point in the direction of the magnetic field.

Magnetic Field at the Center of a Circular Current-Carrying Wire Loop

The magnitude of the magnetic field produced by a circular loop of wire of N loops of radius r, and carrying current I at the center of the loop is

$$B = \frac{\mu_0 NI}{2r} \tag{19.13}$$

The direction of $\vec{B}$ can be obtained conveniently by a right-hand rule that is slightly different, but equivalent to the one given for straight current.

If you curl the fingers of your right hand in the direction of the current, the direction of the magnetic field in the area enclosed by the loop is given by the direction in which the extended thumb is pointing.

The magnetic field produced by a single loop of current is similar to that produced by a short bar magnet. One side of the loop behaves like a magnetic north pole, and the other side like a magnetic south pole.

Magnetic Field in a Current-Carrying Solenoid

A *solenoid* is a long coil of wire wound into a succession of closely spaced loops. The magnetic field at the center of a solenoid of length L having N loops and carrying current I is

$$B = \frac{\mu_0 NI}{L} \tag{19.14}$$

2007 Pearson Education, Inc., Upper Saddle River, NJ. All rights reserved. This material is protected under all copyright laws as they currently exist. No portion of this material may be reproduced, in any form or by any means, without permission in writing from the publisher.

or $B = \mu_0 n I$

where $n = N/L$, the number of turns per meter, called the *linear turn density*.

The magnetic field inside a long solenoid is uniform and directed along the axis of the solenoid. The magnetic field outside the solenoid is nearly zero. The magnetic field lines of a solenoid resemble those of a permanent bar magnet.

19.7 Magnetic Materials

The origin of magnetic fields is circulating electric currents due to the motion of electrons in atoms.

An electron orbiting the nucleus of an atom behaves like a tiny loop of current generating a tiny magnetic field. The spinning motion of an electron also gives rise to a magnetic field. For most substances the net magnetic field due to the orbital and spinning motions of different electrons is nearly zero, and the substance is nonmagnetic.

Ferromagnetic materials have magnetic fields due to the spinning motion of electrons that are nonzero. These materials have small regions, called **magnetic domains**, within the material in which electron spins are naturally parallel to each other. The most common ferromagnetic materials are iron, cobalt, and nickel. Gadolinium and certain alloys are also found to be ferromagnetic. Each domain has a strong magnetic field, but the domains are oriented randomly, and the net magnetic effect is small. When the material is placed in an external magnetic field, the magnetic domains with orientations parallel to the field grow at the expense of other domains, and the orientations

230

© 2007 Pearson Education, Inc., Upper Saddle River, NJ. All rights reserved. This material is protected under all copyright laws as they currently exist. No portion of this material may be reproduced, in any form or by any means, without permission in writing from the publisher.

of other domains become more aligned with the field. They remain aligned for the most part, and the material becomes permanently magnetized.

Electromagnets and Magnetic Permeability
Ferromagnetic materials, such as *soft iron*, are used in electromagnets by wrapping a wire around the core of the ferromagnetic material. The current in the coil generates a magnetic field in the core whose own magnetic field is much larger than the coil's magnetic field. Magnetism can be turned on and off by turning the current on and off. Soft iron is used since in soft iron the magnetic domains quickly become unaligned (that is, it is demagnetizsed) when the current is off (that is, when the external field is removed).

When an electromagnet is on, the iron core provides an additional magnetic field. The total field at the center of an iron-core solenoid is

$$B = \frac{\mu NI}{L} \qquad (19.15)$$

where μ is called the **magnetic permeability** of the core material. For magnetic materials, the magnetic permeability is defined as

$$\mu = \kappa_m \mu_0 \qquad (19.16)$$

where κ_m is called the relative permeabiity. Ferromagnetic materials have values of κ_m on the order of thousands.

Ferromagnets can lose their magnetic field by hitting such a magnet with a hard object or dropping from a height. All ferromagnets lose their magnetic fields if they are heated above a certain critical temperature, called the **Curie temperature**, to orient the atoms in random directions.

231

007 Pearson Education, Inc., Upper Saddle River, NJ. All rights reserved. This material is protected under all copyright laws as they currently exist. No portion of this material may be reproduced, in any form or by any means, without permission in writing from the publisher.

Ferromagnetic domain alignment has applications in geophysics. A typical volcano has many lava flows over time. When the lava cools below the Curie temperature, the ferromagnetic minerals in the lava are aligned in the direction of Earth's magnetic field. Recording the strength and orientation of these older lava flows, geologists determined the changes of Earth's magnetic field and polarity over time. It was found that over a period of millons of years the field, as well as the location of the poles and polarity of Earth's magnetic field, had changed. These changes can provide clues to the *origin* of Earth's magnetic field. The first evidence supporting plate tectonic motion was found from the measurement of the direction of the magnetic polarity of seafloor samples containing iron.

Some living organisms are known to incorporate small ferromagnetic crystals consisting of magnetite in their bodies.

*19.8 Geomagnetism: Earth's Magnetic Field

Like many planets, Earth has its own magnetic field. In ancient times magnetized needles were used by navigators to locate north. Certain bacteria and homing pigeons use Earth's magnetic field for navigation.

In many respects, Earth's magnetic field (also called *geomagnetic field*) is similar to that of a giant bar magnet inside the Earth with its poles near the geographic poles. The geographic North Pole of Earth is actually very close to the south pole of Earth's magnetic field. The magnitude of the horizontal component of Earth's magnetic field at

232

© 2007 Pearson Education, Inc., Upper Saddle River, NJ. All rights reserved. This material is protected under all copyright laws as they currently exist. No portion of this material may be reproduced, in any form or by any means, without permission in writing from the publisher.

the magnetic equator is about 10^{-5} T and the vertical component at the geomagnetic poles is about 10^{-4} T. The origin of Earth's magnetic field is still not completely understood. A permanent solid magnet cannot exist inside the Earth because the temperature at the interior is well above the Curie temperature of a ferromagnetic material. However, it is accepted that the flowing current of molten materials in Earth's core is the primary cause of the magnetic field.

The axis of Earth's magnetic field does not lie along Earth's rotational axis. A compass needle points in the direction of *magnetic north*, not geographic north. The angular difference between the two directions is called the *magnetic declination*. Magnetic declination varies with place and time.

Earth's magnetic poles drift very slowly with time, and they have reversed many times over the ages, most recently about 700 000 years ago. By analyzing the evidence of these reversals in the rock of the ocean floors, geologists found support for the theory of continental drift.

The charged particles emitted by the sun enter Earth's magnetic field at an angle other than 90° sprials in a helix. In a nonuniform bulging magnetic field the particles spiral back and forth as though confined in a magnetic bottle. An analogous phenomenon occurs in Earth's magnetic field, giving rise to two large donut-shaped regions (called *Van Allen radiation belts*) with concentrations of charged particles. In the lower Van Allen belt light emissions called *aurorae* occur.

It is believed that an aurora is created after a violent solar disturbance when incoming charged particles from the sun

233

2007 Pearson Education, Inc., Upper Saddle River, NJ. All rights reserved. This material is protected under all copyright laws as they currently exist. No portion of this material may be reproduced, in any form or by any means, without permission in writing from the publisher.

are trapped in Earth's magnetic field. Near Earth's magnetic poles, they collide with atmospheric atoms and molecules, resulting in the emission of light called aurora borealis near the North Pole and aurora australis near the South Pole.

Hints and Suggestions for Solving Problems

1. The magnetic force on a charge is zero if the charge is at rest. A magnetic force occurs only when a charged particle moves in a direction different from the direction of the magnetic field lines.

2. The magnitude of the magnetic force on a charged particle is given by $F = qvB \sin \theta$. Use this equation to solve problems in Section 19.2.

3. Remember in a velocity selector the net force is zero on a charge when the electric and magnetic forces are equal and opposite. This gives $v = E/B_1$. Also remember that the second magnetic field in a mass spectrometer provides the necessary centripetal force. This gives $qvB_2 = mv^2/r$. These relations will be useful for solving some problems in Section 19.3.

4. Many equations in this chapter depend directly on the charge q on a particle. Make sure that the sign of the result corresponds to the sign on the charge.

5. The magnetic force right-hand rule provides the direction of magnetic force for a positively charged particle. The direction of the force will be the opposite if the charge on the particle is negative.

234

© 2007 Pearson Education, Inc., Upper Saddle River, NJ. All rights reserved. This material is protected under all copyright laws as they currently exist.
No portion of this material may be reproduced, in any form or by any means, without permission in writing from the publisher.

6. To solve problems of magnetic force on a current-carrying wire, such as those in Section 19.4, use the expression $F = ILB \sin \theta$ for the magnetic force. Here θ is the angle between the wire of length L and the magnetic field, B.

7. To calculate torque on current-carrying loops in a magnetic field, such as in the problems in Section 19.5, use the relation $\tau = NIAB \sin \theta$. Recall that here θ is the angle between the magnetic field and the *normal* to the area of the loop. Note that the torque does not depend on the shape of the area. Also, $\tau = mB \sin \theta$ where $m = NLA$, the magnetic moment.

8. The magnetic field inside a solenoid is uniform and parallel to the axis of the solenoid given by $B = \mu_0 nI$. Here n is the number of turns per unit length of the solenoid. For an ideal solenoid the magnetic field outside the solenoid is zero.

2007 Pearson Education, Inc., Upper Saddle River, NJ. All rights reserved. This material is protected under all copyright laws as they currently exist. No portion of this material may be reproduced, in any form or by any means, without permission in writing from the publisher.

CHAPTER 20
ELECTROMAGNETIC INDUCTION
AND WAVES

This chapter describes magnetic flux, Faraday's magnetic induction, and Lenz's law. Electric generators, transformers, and electromagnetic waves are also described.

Key Terms and Concepts

Electromagnetic induction
Induced emf
Magnetic flux
Weber
Faraday's law and Lenz's law of induction
Electric generators
Alternating current and ac generator
Back emf
Transformer
Eddy currents
Electromagnetic waves
Radiation pressure
Radio and TV waves
Infrared radiation
Visible light
Ultraviolet radiation
X-rays
Gamma rays
Computerized tomography

© 2007 Pearson Education, Inc., Upper Saddle River, NJ. All rights reserved. This material is protected under all copyright laws as they currently exist. No portion of this material may be reproduced, in any form or by any means, without permission in writing from the publisher.

20.1 Induced Emf's: Faraday's Law and Lenz's Law

Orsted discovered that an electric current generates a magnetic field. Faraday discovered the reverse effect—that a changing magnetic field generates an electric current.

Faraday found that when a magnet is moved toward or away from a stationary closed wire loop, or when a closed wire loop is moved toward a stationary magnet, the needle of a galvanometer connected in series with the coil deflects. This indicated the presence of a current called *induced current* in the coil. He also noticed that a current is induced in a stationary current wire loop when the current in another loop close to it is varied. In these situations, the current induced in a loop is caused by an induced electromotive force (emf) due to a process called **electromagnetic induction**. Faraday concluded from his observations that:

- Whenever the number of magnetic field lines changes in a closed coil, an induced current is developed in the coil.

- The magnitude of the induced current is proportional to the rate of change of the magnetic field, and its direction depends on the direction of the change of the magnetic field.

Magnetic Flux
Magnetic flux is a measure of the number of magnetic field lines that cross a given area.

237

©2007 Pearson Education, Inc., Upper Saddle River, NJ. All rights reserved. This material is protected under all copyright laws as they currently exist. No portion of this material may be reproduced, in any form or by any means, without permission in writing from the publisher.

If the magnetic field (B) lines cross an area A at an angle θ with the normal to the area, the magnetic flux through the area is

$$\phi = BA \cos \theta \qquad (20.1)$$

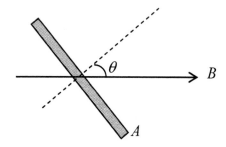

The SI unit of magnetic flux is tesla-meter square or weber (Wb). 1 Wb = 1 T·m²

Note that:

ϕ is maximum ($\phi_{max} = BA$), when $\theta = 0°$.

ϕ is maximum with opposite sign ($\phi_{180} = -BA$), when $\theta = 180°$.

$\phi = 0$ when $\theta = 90°$.

For intermediate angles, the flux is less than the maximum value.

Faraday's Law of Induction and Lenz's Law

The basic feature of magnetic induction can be expressed mathematically using the concept of magnetic flux. Whenever magnetic flux through a closed coil changes, an emf is induced in the coil. The *induced emf depends on the time rate of change of magnetic flux*, $\Delta\phi/\Delta t$.

Faraday's law of induction:

$$\varepsilon = -N \frac{\Delta\phi}{\Delta t} = -\frac{\Delta(N\phi)}{\Delta t} \qquad (20.2)$$

238

© 2007 Pearson Education, Inc., Upper Saddle River, NJ. All rights reserved. This material is protected under all copyright laws as they currently exist. No portion of this material may be reproduced, in any form or by any means, without permission in writing from the publisher.

where N is the number of loops in the coil. The negative sign in front of N indicates the *polarity* or *direction* of the emf. The direction is found by **Lenz's law**:

An induced emf gives rise to a current whose magnetic field opposes the change in magnetic flux that produced it.

For example, if the magnetic flux increases in the $+x$ direction, the magnetic field due to the induced current will be in the $-x$ direction. The induced current direction is given by the **induced-current right-hand rule**: *With the fingers of the right hand pointing in the direction of the induced field, the extended thumb points in the direction of the induced current.*

Lenz's law is a consequence of the law of conservation of energy. If Lenz's law were not true, that is if the magnetic field of the induced emf would add to the original field, it would create energy out of nothing, violating conservation of energy.

In most situations, the magnitude of the induced emf is calculated by using Faraday's law, and its direction is determined by using Lenz's law.

From Equation (20.1) and Equation (20.2)

$$\varepsilon = -N\,\frac{\Delta\phi}{\Delta t} = -N\,\frac{\Delta\left(BA\cos\theta\right)}{\Delta t} \qquad (20.3)$$

Clearly an induced emf is produced when the magnetic field B changes and/or the area A changes and/or the angle between the normal to the loop area and the field changes.

239

2007 Pearson Education, Inc., Upper Saddle River, NJ. All rights reserved. This material is protected under all copyright laws as they currently exist. No portion of this material may be reproduced, in any form or by any means, without permission in writing from the publisher.

20.2 Electric Generators and Back Emf

Electric Generators

An *electric generator* converts mechanical energy to electrical energy. In an electric generator a closed coil is rotated in a constant magnetic field. As the coil rotates, it changes the magnetic flux crossing the coil. As a result an induced emf is generated in the coil based on Faraday's law of magnetic induction.

The induced emf produced by a generator is

$$\mathcal{E} = -N\frac{\Delta\phi}{\Delta t} = -N\frac{\Delta(BA\cos\theta)}{\Delta t}$$

$$= -NBA\frac{\Delta(\cos\theta)}{\Delta t} = NBA\omega\sin\omega t$$

$$\mathcal{E} = \mathcal{E}_0\sin\omega t \qquad (20.4)$$

where $\mathcal{E}_0 = NBA\omega$ is the maximum magnitude of the emf.

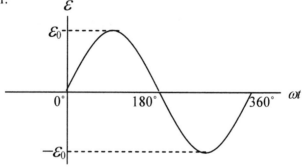

If f is the rotational frequency of the generator's armature, $\omega = 2\pi f$. Thus, from Equation (20.4):

$$\mathcal{E} = \mathcal{E}_0\sin 2\pi ft \qquad (20.5)$$

240

© 2007 Pearson Education, Inc., Upper Saddle River, NJ. All rights reserved. This material is protected under all copyright laws as they currently exist. No portion of this material may be reproduced, in any form or by any means, without permission in writing from the publisher.

It is clear from the preceding expression that the induced emf, and hence the induced current, changes its magnitude and sign (from $\mathcal{E}_0$ to $-\mathcal{E}_0$). This type of current is called *alternating current*. The plot of $\mathcal{E}$ is shown.

Back Emf

A motor is a generator run in reverse. Like a generator, a motor has a rotating armature in a magnetic field. As a result, it generates an induced emf. This induced emf is called *back emf* ($\mathcal{E}_b$) since its polarity is opposite to that of the line voltage that drives the motor. If the coil of the motor has an internal resistance, R, and if I is the current the motor draws when in operation, it can be shown that the back emf is given by

$$\mathcal{E}_b = V - IR \qquad (20.6)$$

where V is the *line* voltage.

The back emf depends on the rotational speed of the armature and builds up from zero to some maximum value as the armature goes from rest to its normal operating speed. Back emf also develops in a generator. When an operating generator is not connected to a load, there is no current and, hence, no magnetic force on the coils of the armature. When the generator delivers a current to a load, because of the current flow in the coil, the coil exerts a magnetic force that produces a counter torque that opposes the rotation of the torque.

2007 Pearson Education, Inc., Upper Saddle River, NJ. All rights reserved. This material is protected under all copyright laws as they currently exist. No portion of this material may be reproduced, in any form or by any means, without permission in writing from the publisher.

20.3 Transformers and Power Transmission

A transformer is an electrical device that can increase or decrease an ac voltage keeping electric *power* essentially unchanged. It works on the principle of Faraday's magnetic induction.

A simple **transformer** consists of an iron core on which two coils of insulated wire are wound: a *primary coil* with N_p number of turns and a *secondary coil* with N_s number of turns. The primary coil is connected to the ac voltage that needs to be increased or decreased. The sSecondary coil is connected to the load.

As the alternating current in the primary coil changes, the flux through the primary and secondary coils also changes. The following transformer equations give the relationship between current, I, voltage, V, and number of turns, N, of the primary and secondary circuits:

$$\frac{V_s}{V_p} = \frac{N_s}{N_p} \tag{20.7}$$

$$I_p V_p = I_s V_s \tag{20.8}$$

$$\frac{I_p}{I_s} = \frac{V_s}{V_p} = \frac{N_s}{N_p} \tag{20.9}$$

Thus, the output voltage and output current are

© 2007 Pearson Education, Inc., Upper Saddle River, NJ. All rights reserved. This material is protected under all copyright laws as they currently exist. No portion of this material may be reproduced, in any form or by any means, without permission in writing from the publisher.

$$Vs = \left(\frac{N_s}{N_p} \right) V_p \qquad (10.10a)$$

and

$$Is = \left(\frac{N_p}{N_s} \right) I_p \qquad (10.10b)$$

If $N_s > N_p$, $V_s > V_p$, the voltage is stepped up, and the transformer is called a *step-up transformer*.

If $N_s < N_p$, $V_s < V_p$, the voltage is stepped down, and the transformer is called a *step-down transformer*. Clearly, a step-up transformer can be used as a step-down transformer by reversing the output and input connections.

In actual transformers there are some energy losses. There is always some flux leakage. Also, when ac current flows in the primary, the changing magnetic flux through the loops gives rise to an induced emf in the coil that opposes the change of current. This phenomenon is called *self induction*. The core of a transformer is made of highly permeable materials (that are good conductors) to increase the density of magnetic flux through the core. The changing magnetic flux in the core sets up the motion of charges in the core, generating a current, called *eddy currents*. To reduce eddy current losses, the cores are usually made of thin sheets of material laminated with an insulating glue between them. Internal energy loss is less than 5% in well-designed transformers.

Power Transmission and Transformers
For power transmission over long distances, the output voltage from a generator is stepped up and transmitted

243

2007 Pearson Education, Inc., Upper Saddle River, NJ. All rights reserved. This material is protected under all copyright laws as they currently exist. No portion of this material may be reproduced, in any form or by any means, without permission in writing from the publisher.

over long distances to an area substation near the consumer. Because the voltage is stepped up, the current of the generator's output is reduced and, as a result, I^2R losses during transmission are reduced. Voltage is stepped down at the distributing substations before being supplied to homes and businesses.

20.4 Electromagnetic Waves

Electromagnetic waves are traveling waves of oscillating electric and magnetic fields. The electric and magnetic fields in electromagnetic waves are perpendicular to each other, and they are also perpendicular to the direction of propagation of the waves as shown. Clearly, electromagnetic waves are transverse. Both the electric and magnetic fields carry energy and propagate outward with the speed of light, c ($c = 3.00 \times 10^8$ m/s in vacuum).

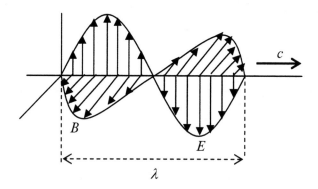

Electromagnetic waves can be produced using an ac electric circuit and an antenna. An accelerated charge radiates energy in the form of electromagnetic waves. Maxwell showed that four fundamental relations, called

© 2007 Pearson Education, Inc., Upper Saddle River, NJ. All rights reserved. This material is protected under all copyright laws as they currently exist. No portion of this material may be reproduced, in any form or by any means, without permission in writing from the publisher.

Maxwell equations, completely describe all observed electromagnetic phenomena.

Radiation Pressure

Since an electromagnetic wave carries energy, it can exert a force on a material it strikes. The electric field of the wave acts on an electron of the surface, giving it velocity. The magnetic field then exerts a force on the moving electron in the direction of propagation of the wave. The radiation force per unit area is called **radiation pressure**. Radiation pressure is usually very small but is important in atmospheric and astronomical phenomena. For example, as a comet approaches the sun, the tail of the comet always points away from the sun because of the solar radiation pressure.

Types of Electromagnetic Waves

The speed, c, of an electromagnetic wave is related to its frequency, f, and its wavelength, λ, by $c = f\lambda$. The ordered series containing electromagnetic waves of increasing frequency or wavelength is called the electromagnetic spectrum. The spectrum is continuous.

Although the boundary between different regions is not sharp, several bands of frequencies in the spectrum were given the following names:

Power Waves: Electromagnetic waves of frequency 60 Hz are produced in alternating electric circuits. This power wave has a very long wavelength of 5.00×10^6 m.

2007 Pearson Education, Inc., Upper Saddle River, NJ. All rights reserved. This material is protected under all copyright laws as they currently exist. No portion of this material may be reproduced, in any form or by any means, without permission in writing from the publisher.

Type of Wave	Approximate Frequency Range (Hz)
Power waves	60
Radio waves	
AM	$(0.53 \times 10^6) - (1.7 \times 10^6)$
FM	$(88 \times 10^6) - (108 \times 10^6)$
TV	$(543 \times 10^6) - (890 \times 10^6)$
Microwaves	$10^9 - 10^{11}$
Infrared	$10^{11} - 10^{14}$
Visible light	$4.0 \times 10^{14} - 7.0 \times 10^{14}$
Ultraviolet	$10^{15} - 10^{17}$
X-rays	$10^{17} - 10^{19}$
Gamma rays	Greater than 10^{19}

Radio and TV Waves: Radio and TV waves are in the frequency range between 500 kHz and 1000 MHz. The AM (amplitude-modulated) band runs between 530 and 1710 kHz, and frequencies up to 54 MHz are used for shortwave bands. TV waves run between 54 MHz and 890 MHz. The FM (frequency-modulated) radio band runs between 88 and 108 MHz. Cellular phones use eletromagnetic waves in the ultrahigh frequency band. Radio waves are reflected by the ionic layers in the upper atmosphere, thus making the transmission possible around the curvature of the Earth.

Microwaves: Special vacuum tubes called klystrons and magnetrons produce microwaves, which are in the gigahertz frequency range. Microwaves are used in microwave ovens. Microwaves are also used in communications and radar applications.

Infrared Radiation: Infrared radiation has longer wavelength or shorter frequency than visible light. Any object at room temperature primarily emits infrared radiation. Infrared raditions

246

© 2007 Pearson Education, Inc., Upper Saddle River, NJ. All rights reserved. This material is protected under all copyright laws as they currently exist No portion of this material may be reproduced, in any form or by any means, without permission in writing from the publisher.

e also called heat rays. Infrared lamps are used to keep food arm.

isible Light: Visible light has wavelengths between 400 nm ad 700 nm or the frequencies between 4×10^{14} Hz and 7×10^{14} z. Human eyes are sensitive to visible light.

ltraviolet Radiation: Ultraviolet radiation has a shorter avelength than visible light. Special lamps and very hot bjects produce it. The sun emits large amounts of ultraviolet diation, most of which is absorbed by the ozone layer at an titude of about 40 km from the surface of the Earth. Most traviolet radiation is absorbed by ordinary glass.

-Rays: X-rays have shorter wavelengths (that is, higher equencies) than ultraviolet radiation. In an X-ray tube ectrons are accelerated through a large voltage and strike a rget electrode. There they decelerate and excite the atomic ectrons of the target. X-ray radiation is emitted during eceleration and as the atoms are de-excited. X-rays are used to ke photographs of bones. High-frequency X-rays have high ergy and cause cancer, skin burns, and other harmful effects. omputerized X-ray machines can take three-dimensional nages by means of a technique called *computerized mography.*

amma Rays: Gamma rays are the most energetic among all ectromagnetic waves and have the highest frequency in the ectromagnetic spectrum. These are produced in nuclear actions, radioactivity, and in particle accelerators.

2007 Pearson Education, Inc., Upper Saddle River, NJ. All rights reserved. This material is protected under all copyright laws as they currently exist. No portion of this material may be reproduced, in any form or by any means, without permission in writing from the publisher.

Hints and Suggestions for Solving Problems

1. Use the expression $\phi = BA \cos \theta$ to calculate magnetic flux passing through an area A, where θ is the angle between the magnetic field B and the *normal* to A.

2. Remember that the magnetic flux through a coil can be changed by changing the magnetic field, by changing the area, by changing angle θ, or by changing all.

3. Use Faraday's law, $\mathcal{E} = -N \dfrac{\Delta \phi}{\Delta t}$ to determine the induced emf, and then use Ohm's law to determine the induced current from the induced emf.

4. The direction of the induced emf follows from Lenz's law. Lenz's law is simple but often confusing. Practice by applying it to different problems.

5. For a rotating coil, such as in a generator (see problems in Section 20.2), induced emf can be obtained from $\mathcal{E} = (NBA\omega) \sin \omega t$.

6. For a transformer, the voltage ratio (secondary to primary) is the same as the turns ratio (secondary to primary) of the coils. The current ratio is the reciprocal of the voltage ratio. This information can be used to solve problems in Section 20.3.

7. Note that all electromagnetic waves travel with the same speed, which is the speed of light, c. $c = f\lambda$, where $c = 3.00 \times 10^8$ m/s in vacuum. Use this information to solve problems in Section 20.4.

© 2007 Pearson Education, Inc., Upper Saddle River, NJ. All rights reserved. This material is protected under all copyright laws as they currently exist. No portion of this material may be reproduced, in any form or by any means, without permission in writing from the publisher.

CHAPTER 21
AC CIRCUITS

This chapter describes alternating-current circuits
involving resistors, capacitors, and inductors. It discusses
impedance, phase difference, phasor diagram, ac power,
power factors, and resonance in ac electric circuits.

Key Terms and Concepts

> Alternating voltage
> Alternating current
> Peak voltage and peak current
> RMS current or effective current
> RMS voltage or effective voltage
> Capacitive reactance
> Inductance
> Inductor
> Inductive reactance
> Phase diagram
> Phasors
> Impedance
> Power factor
> Series *RLC* circuit
> Resonance in electric circuit
> Resonant frequency

Although direct (dc) currents have many uses, most of the
electrical devices that we use require alternating currents
(ac). The generators produce ac power, transformers can
change the voltage and current in ac circuits, and ac power
can be transmitted economically over very long distances.
These are the advantages of ac over dc currents.

2007 Pearson Education, Inc., Upper Saddle River, NJ. All rights reserved. This material is protected under all copyright laws as they currently exist.
No portion of this material may be reproduced, in any form or by any means, without permission in writing from the publisher.

Whereas in a common dc circuit only ohmic resistance plays the role of controlling current, in ac circuits, in addition to resistors, capacitors and inductors also play the same role.

21.1 Resistance in an AC Circuit

A circuit containing an ac voltage source and a resistance, R, is shown.

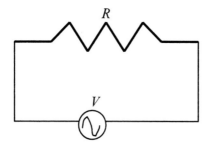

The variation of ac voltage with time is expressed mathematically as

$$V = V_0 \sin \omega t = V_0 \sin 2\pi f t \qquad (21.1)$$

where V_0 is the maximum value of the voltage, called the **peak voltage**, and ω is the angular frequency ($\omega = 2\pi f$).

AC Current and Power
The resulting current, I, flowing through the resistor, R, is

$$I = \frac{V}{R} = \frac{V_0}{R} \sin 2\pi f t$$

or $\qquad I = I_0 \sin 2\pi f t \qquad (21.2)$

250

© 2007 Pearson Education, Inc., Upper Saddle River, NJ. All rights reserved. This material is protected under all copyright laws as they currently exis
No portion of this material may be reproduced, in any form or by any means, without permission in writing from the publisher.

where $I_0 = V_0/R$, called the **peak current**.

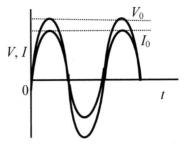

The voltage and the current have the same time variation, and they are *in phase* (that is, zero values and maxima occurring at the same times) with each other.

The average value of the current in one cycle is zero, since average of sin θ is zero for a cycle.

The instantaneous power is given by
$$P = I^2 R = I_0^2 R \sin^2 2\pi ft \qquad (21.3)$$
Since the average value of $\sin^2 2\pi ft$ is ½,

$$\bar{I}^2 = \frac{1}{2} I_0^2 \qquad (21.4)$$

Thus, the average power is

$$\bar{P} = \frac{1}{2} I_0^2 R \qquad (21.5)$$

It is customary to write ac power in the form $I^2 R$, the dc power. In order to do so, we define

$$I_{\text{rms}} = \frac{1}{\sqrt{2}} I_0 = 0.707 \, I_0 \qquad (21.6)$$

where I_{rms} is called the root mean square current, commonly known as **rms current** or **effective current**. Thus the average power is

251

2007 Pearson Education, Inc., Upper Saddle River, NJ. All rights reserved. This material is protected under all copyright laws as they currently exist.
No portion of this material may be reproduced, in any form or by any means, without permission in writing from the publisher.

$$\overline{P} = I_{rms}^2 R \qquad (21.7)$$

AC Voltage

The **rms voltage** or **effective voltage** is defined as

$$V_{rms} = \frac{1}{\sqrt{2}} V_0 = 0.707 \, V_0 \qquad (21.8)$$

Since the peak values of the voltage and current are related by $V_0 = I_0 R$, clearly,

$$V_{rms} = I_{rms} R \qquad (21.9)$$

The average power, thus, can be written as

$$\overline{P} = I_{rms}^2 R = I_{rms} V_{rms} = \frac{V_{rms}^2}{R} \qquad (21.10)$$

The rms values are usually measured or specified for ac quantities. For example, ac ammeters and voltmeters provide readings of rms values of current and voltage, respectively. The line voltage 120 V is rms voltage, whose peak value is 1.414 (120 V) = 170 V. The *peak-to-peak* voltage V_{p-p} is defined as twice the peak voltage, that is, $V_{p-p} = 2V_0$.

21.2 Capacitive Reactance

When a capacitor is connected with an ac voltage source, the capacitor is alternately charged and discharged as the voltage reverses each half-cycle. When the voltage is maximum, the capacitor is fully charged ($Q_0 = CV_0$). Since the plate cannot accumulate any more charge, there is no current in the circuit at this instant. As the voltage decreases to zero from its maximum value, the capacitor discharges, giving rise to a current that becomes maximum when the voltage is zero. As the voltage then increases in magnitude in the opposite direction, the capacitor charges,

© 2007 Pearson Education, Inc., Upper Saddle River, NJ. All rights reserved. This material is protected under all copyright laws as they currently exist. No portion of this material may be reproduced, in any form or by any means, without permission in writing from the publisher.

giving rise to current that decreases, becoming zero halfway through the cycle. During the next half-cycle, the process is reversed. Clearly, the current and voltage are not in step or in phase. *In a purely capacitive ac circuit, the current leads the voltage by 90°, or by one-quarter cycle.*

A capacitor in an ac circuit provides opposition to the charging process. This opposition is expressed by **capacitive reactance** (X_C), given by

$$X_C = \frac{1}{2\pi f C} = \frac{1}{\omega C} \qquad (21.11)$$

The SI unit of capacitive reactance is the ohm, or second per farad (s/F).

Capacitive reactance decreases as frequency increases. At very high frequency, voltage changes direction so rapidly that there is not enough time to charge the capacitor. As a result, a capacitor offers essentially no resistance to the flow of charges onto its plates. On the other hand, if $f = 0$, capacitive reactance is infinite.

The current-voltage relation for the capacitor connected to an ac source is

$$V_{rms} = I_{rms} X_C \qquad (21.12)$$

where V_{rms} and I_{rms} represent the rms values of voltage and current, respectively.

©2007 Pearson Education, Inc., Upper Saddle River, NJ. All rights reserved. This material is protected under all copyright laws as they currently exist. No portion of this material may be reproduced, in any form or by any means, without permission in writing from the publisher.

21.3 Inductive Reactance

Inductance is an ability of a circuit element to oppose a time-varying current. When placed in a circuit, such an element, called an **inductor**, will generate reverse voltage in opposition to the time-varying current.

The changing current through an inductor produces a changing magnetic flux through the inductor. As a result, a back emf is induced in the coil by Faraday's magnetic induction. This phenomenon is called *self induction*.

According to Faraday's law of magnetic induction, the self-induced emf is given by

$$\mathcal{E} = -N \frac{\Delta\phi}{\Delta t}, \text{ which can be written as}$$

$$\mathcal{E} = -L \frac{\Delta I}{\Delta t} \qquad (21.13)$$

since the time rate of change of flux $\frac{\Delta\phi}{\Delta t}$ is proportional to the time rate of change of current, $\frac{\Delta I}{\Delta t}$. The constant L in the preceding equation is called the inductance of the coil. Its unit is volt·second per ampere (V·S/A), called **henry** (H). 1 millihenry (mH) = 10^{-3} henry (H).

The opposition presented by an inductor in an ac circuit is directly proportional to the frequency of the ac and is given by the **inductive reactance** (X_L):

$$X_L = 2\pi f L = \omega L \qquad (21.14)$$

X_L increases with frequency since, as the frequency increases, the current changes more rapidly with time, thus generating greater emf across the inductor.

254

© 2007 Pearson Eductance, Inc., Upper Saddle River, NJ. All rights reserved. This material is protected under all copyright laws as they currently exist. No portion of this material may be reproduced, in any form or by any means, without permission in writing from the publisher.

The SI unit of X_L is ohm (Ω) or henry per second (H/s). The voltage across an inductor connected to an ac source can be written as

$$V_{rms} = I_{rms} X_L \qquad (21.15)$$

When an inductor is connected to an ac source, the maximum voltage corresponds to zero current, and when the voltage drops to zero, the current is a maximum. This occurs because when the voltage changes its polarity, the inductor tries to prevent this change, generating an opposing induced emf and current. Thus, *in a purely inductive ac circuit, the voltage leads the current by 90°, or one-quarter cycle.*

21.4 Impedance: *RLC* Circuits

A real ac circuit usually has resistance and either or both a capacitor and inductor.

Series *RC* Circuit

In a circuit consisting of a resistor, R, and a capacitor, C, in series with an ac voltage source, the phase difference between the current and voltage is different for each of R and C. A special graphical method called a *phase diagram* is used to find the effective opposition to the current. In a **phase diagram**, the resistance and reactances are treated as vectors, and their magnitudes, represented as arrows, are called **phasors**. The vector sum is the effective opposition to the current, called the **impedance** (Z).

2007 Pearson Education, Inc., Upper Saddle River, NJ. All rights reserved. This material is protected under all copyright laws as they currently exist. No portion of this material may be reproduced, in any form or by any means, without permission in writing from the publisher.

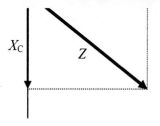

From the preceding phase diagram,

$$Z = \sqrt{R^2 + X_C^2} \qquad (21.16)$$

Clearly, the impedance of an ac circuit is analogous to resistance in a dc circuit. For an *RC* circuit, the impedance takes into account both the resistance, R, and the capacitive reactance, X_C.

The relationship between voltage and current for a circuit containing R, L, and C is given by the generalized Ohm's law:

$$V_{rms} = I_{rms}Z \qquad (21.17)$$

The SI unit of impedance is the ohm.

Series *RL* Circuit

In a series *RL* circuit, the impedance Z is the phasor sum of the resistance R and the inductive reactance X_L, as shown in the phase diagram below. Here X_L is plotted along the $+y$ axis to indicate a phase difference of $+90°$.

© 2007 Pearson Education, Inc., Upper Saddle River, NJ. All rights reserved. This material is protected under all copyright laws as they currently exist. No portion of this material may be reproduced, in any form or by any means, without permission in writing from the publisher.

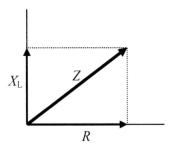

The impedance, $Z = \sqrt{R^2 + X_L^2}$ (21.18)

Series *RLC* Circuits

A circuit consisting of a resistor, R, an inductor, L, and a capacitor, C, in series with an ac generator is known as an *RLC* circuit.

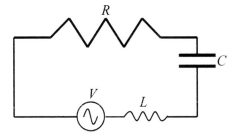

The behavior of an *RLC* circuit can be analyzed using a phasor diagram. The voltage of the resistor is in phase with the current, whereas the voltage of the inductor and capacitor point in opposite directions (90° ahead of the current in the inductor and 90° behind the current for the capacitor), as shown in the phase diagram.

257

2007 Pearson Education, Inc., Upper Saddle River, NJ. All rights reserved. This material is protected under all copyright laws as they currently exist. No portion of this material may be reproduced, in any form or by any means, without permission in writing from the publisher.

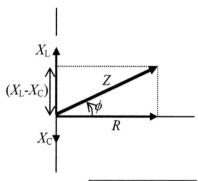

The impedance is $Z = \sqrt{R^2 + \left(X_L - X_C\right)^2}$ (21.19)

The phase angle (ϕ) between the voltage and current is

$$\tan \phi = \frac{X_L - X_C}{R} \qquad (21.20)$$

If $X_L > X_C$, $\phi > 0$, the voltage leads the current, and the circuit is called an *inductive* circuit.

If $X_L < X_C$, $\phi < 0$, the voltage lags the current, and the circuit is called a *capacitive* circuit.

If $X_L = X_C$, $\phi = 0$, and the voltage and the current are in phase.

Power Factor for a Series *RLC* Circuit
There are no power losses in a pure capacitor and inductance. Capacitors and inductors store energy and then give it back. The only element that dissipates power in a series *RLC* circuit is the resistor. Thus, the average power dissipated is

$$\overline{P} = P_R = I_{rms}V_R$$

© 2007 Pearson Education, Inc., Upper Saddle River, NJ. All rights reserved. This material is protected under all copyright laws as they currently exist. No portion of this material may be reproduced, in any form or by any means, without permission in writing from the publisher.

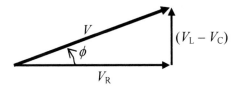

From the preceding diagram, $V_R = V_{rms}\cos\phi$ (21. 21)
and from the phase diagram

$$\cos\phi = \frac{R}{Z} \tag{21.22}$$

Thus, $\overline{P} = I_{rms}V_{rms}\cos\phi$ (21.23)

$$= \hat{I}_{rms}^2 Z\cos\phi \tag{21.24}$$

When $\phi = 0$, $Z = R$, the power dissipated is $P = I_{rms}V_{rms}$ and the circuit is said to be *completely resistive*. In this situation, power dissipation is a maximum.

As ϕ increases in either direction, power decreases. At $\phi = +90°$, the circuit is *completely inductive*, and at $\phi = -90°$, the circuit is *completely capacitive*. Clearly, in these situations, no power is dissipated in the circuit.

21.5 Circuit Resonance

In a series RLC ac circuit, when the power factor $\cos\phi$ is 1, the impedance $Z = R$. In this case, the impedance of the circuit is a minimum and, thus, the current in the circuit is a maximum. This situation is called resonance in the circuit. At resonance, the inductive reactance X_L is the same in magnitude as the capacitive inductance, X_C. The current is in phase with the applied voltage.

259

2007 Pearson Education, Inc., Upper Saddle River, NJ. All rights reserved. This material is protected under all copyright laws as they currently exist. No portion of this material may be reproduced, in any form or by any means, without permission in writing from the publisher.

Since $X_L = X_C$, this gives,

$$2\pi f_0 L = \frac{1}{2\pi f_0 C}$$

Or, $\qquad f_0 = \frac{1}{2\pi\sqrt{LC}}$ $\qquad\qquad$ (21.25)

This frequency f_0, at which the impedance is a minimum or the current is a maximum, is called the **resonant frequency** of a series *RLC* circuit. A plot of current with frequency is shown.

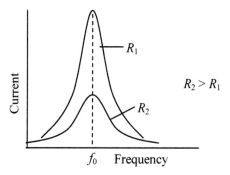

A Physical Explanation of Resonance

The voltage across R is in phase with the current. The voltage across the capacitor lags current by 90°, whereas the voltage across the inductor leads current by 90°. That is, the capacitor and inductor voltages are always 180° out of phase. At resonance, these voltages are equal in magnitude and they cancel exactly. The full voltage appears across R and in this case current is a maximum.

Applications of Resonance

At resonance in a series *RLC* circuit, $Z = R$, $\cos\phi = 1$. In this situation the power dissipation, and hence the current, is a maximum in the circuit. Resonant circuits have

260

© 2007 Pearson Education, Inc., Upper Saddle River, NJ. All rights reserved. This material is protected under all copyright laws as they currently exist.
No portion of this material may be reproduced, in any form or by any means, without permission in writing from the publisher.

different applications. Each radio broadcasting station transmits a unique broadcast frequency. The receiver radio can vary the resonant frequency of the receiving circuit using variable air capacitors. Thus, it selectively picks up only the one whose frequency is the same as or close to the resonant frequency of the receiver circuit.

Hints and Suggestions for Solving Problems

1. Read the problem carefully to find out whether the given currents and voltages are their maximum values or their rms values. If the maximum values are given, the rms values can be obtained by dividing by $\sqrt{2}$. If the rms values are given, maximum values can be obtained by multiplying by $\sqrt{2}$.

2. To calculate the average power, use $\overline{P} = I_{rms}^2 R = V_{rms}^2/R$. To find maximum power, use the maximum values of I and V.

3. It may be useful to draw phasors when solving problems. Draw a capacitor phasor at 90° *clockwise* with respect to the current phasor. Draw an inductor phasor at 90° *counterclockwise* with respect to the current phasor.

4. The rms voltage across a capacitor is $V_{rms} = I_{rms} X_C$, where $X_C = 1/\omega C$. Usually C is given in microfarads. Be sure to convert the value to farads. Also, in most situations linear frequency, f, is given in hertz. Remember to multiply the value by 2π to get ω. The unit of ω is rad/s.

261

2007 Pearson Education, Inc., Upper Saddle River, NJ. All rights reserved. This material is protected under all copyright laws as they currently exist. No portion of this material may be reproduced, in any form or by any means, without permission in writing from the publisher.

5. For an ac circuit, $V_{rms} = I_{rms} Z$. For an RC circuit, the impedance is $Z = \sqrt{R^2 + X_C^2} = \sqrt{R^2 + (1/\omega C)^2}$. For an RL circuit the impedance is $Z = \sqrt{R^2 + X_C^2} = \sqrt{R^2 + (\omega L)^2}$. These expressions will be useful for solving problems in Sections 21.2 - 21.3.

6. For an RLC circuit $Z = \sqrt{R^2 + (X_L - X_C)^2}$. The impedance arises from the resistor, R, the inductive reactance, X_L, and the capacitive reactance, X_C. The angle between the voltage and current is given by $\tan \phi = \dfrac{X_L - X_C}{R}$. Here, ϕ can be greater than, equal to, or less than zero, depending on the values of X_L and X_C. The power factor is $\cos \phi = R/Z$. The preceding expressions will be useful for solving problems in Section 21.4.

7. Remember that at resonance in an RLC circuit, the current and voltage are out-of-phase, and $X_L = X_C$. As a result, $Z = R$. The resonant frequency is $f_0 = 1/2\pi\sqrt{LC}$. It depends on L and C only.

8. To calculate power or the power factor using the expression $P_{av} = I_{rms}V_{rms}\cos\phi$, note that here ϕ is the angle between the voltage and the current phasors, not the angle between the voltage phasor and the x- or y-axis. Also, $\cos\phi = R/Z$.

© 2007 Pearson Education, Inc., Upper Saddle River, NJ. All rights reserved. This material is protected under all copyright laws as they currently exist. No portion of this material may be reproduced, in any form or by any means, without permission in writing from the publisher.

CHAPTER 22
REFLECTION AND REFRACTION
OF LIGHT

This chapter discusses the basics of geometric optics including reflection and refraction of light, total internal reflection, and dispersion. The laws of reflection and refraction are discussed.

Key Terms and Concepts

Reflection of light
Wavefront
Regular or specular reflection
Law of reflection
Irregular or diffuse reflection
Refraction
Huygens' principle
Snell's law
Total internal reflection
Critical angle
Fiber optics
Dispersion

22.1 Wave Fronts and Rays

A line or surface drawn at a particular time through all points of a wave having the same phase is called a **wave front**. For a three-dimensional spherical wave (such as a light wave emitted from a point source) the wavefront is a spherical surface. At very large distances the curvature of a short segment of a spherical wave front is very small. Such a wave front is called a **plane wave front**. A plane wave front can also be produced from a linear, elongated

263

2007 Pearson Education, Inc., Upper Saddle River, NJ. All rights reserved. This material is protected under all copyright laws as they currently exist. No portion of this material may be reproduced, in any form or by any means, without permission in writing from the publisher.

source (such as from a long lightbulb filament). In a uniform medium, the wave fronts move outward at a wave speed. For light the speed in vacuum is $c = 3 \times 10^8$ m/s. The direction of motion of waves is indicated by outward-pointing arrows, called **rays**. Rays and wave fronts are always perpendicular to each other. **Geometrical optics** uses the geometric representations of wave fronts and rays to explain reflection and refraction.

22.2 Reflection

When rays of light strike a surface and return to the same medium, it is called **reflection** of light. The direction of light is changed by reflection. Reflection involves the absorption and re-emission of light by means of complex electromagnetic vibrations of the atoms of the reflecting medium.

When a light wave is reflected by a smooth surface such as by a mirror, the **angle of incidence,** θ_i, is equal to the **angle of reflection,** θ_r. Both angles are measured between the ray and the normal to the surface at the point of incidence. This is called the **law of reflection**:

$$\theta_i = \theta_r \qquad (22.1)$$

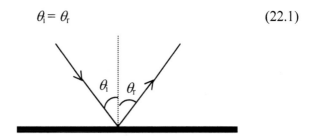

The incident ray, reflected ray, and the normal at the point of incidence lie in the same plane.

264

© 2007 Pearson Education, Inc., Upper Saddle River, NJ. All rights reserved. This material is protected under all copyright laws as they currently exist
No portion of this material may be reproduced, in any form or by any means, without permission in writing from the publisher.

When parallel rays are reflected by a smooth surface, the reflected rays are also parallel to one another. This kind of reflection is called **regular** or **specular reflection**. If the surface is rough, the reflected rays move in different directions because of the irregular nature of the surface. This kind of reflection is called **diffuse** or **irregular reflection**. We can see an illuminated object, such as the moon, because of diffuse reflection from its surface.

If the dimensions of the surface irregularities are greater than the wavelength of light, the reflection is diffuse. Thus, for a good mirror, the surface should be polished until the surface irregularities are about 10^{-7} m, the wavelength of visible light.

22.3 Refraction

Refraction is the change of direction of light at the boundary as it moves from one medium to another. In general, when light is incident on a boundary between two media, some light is reflected and some is refracted. Light refracts because of its change in speed in the second medium.

The change in the direction of light is described by the **angle of refraction**, which is the angle made by the refracted ray with the normal at the point of incidence. If θ_1 is the angle of incidence and θ_2 is the angle of refraction, it can be shown that

$$\frac{\sin \theta_1}{\sin \theta_2} = \frac{v_1}{v_2} \qquad (22.2)$$

007 Pearson Education, Inc., Upper Saddle River, NJ. All rights reserved. This material is protected under all copyright laws as they currently exist. No portion of this material may be reproduced, in any form or by any means, without permission in writing from the publisher.

where v_1 and v_2 are the speed of light in medium 1 and medium 2, respectively. This relation is known as **Snell's law**.

It is customary to compare the speed of light in different media with the speed of light in a vacuum (c) in terms of a ratio, called the **index of refraction** (n):

$$n = \frac{c}{v} = \frac{\text{speed of light in a vacuum}}{\text{speed of light in a medium}} \qquad (22.3)$$

The frequency of light does not change as it passes from one medium to another. Thus,

$$n = \frac{c}{v} = \frac{\lambda}{\lambda_m} \qquad (22.4)$$

λ_m is the wavelength of light in a medium. Clearly,

$$\frac{v_1}{v_2} = \frac{n_2}{n_1}$$

Thus, from Equation (22.2),

$$n_1 \sin \theta_1 = n_2 \sin \theta_2 \qquad (22.5)$$

This is another form of Snell's law.

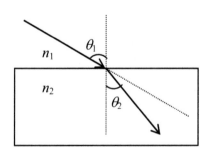

© 2007 Pearson Education, Inc., Upper Saddle River, NJ. All rights reserved. This material is protected under all copyright laws as they currently exist. No portion of this material may be reproduced, in any form or by any means, without permission in writing from the publisher.

Note that in refraction:

if $n_2 > n_1$, the refracted ray bends toward the normal, and

if $n_2 < n_1$, the refracted ray bends away from the normal.

Refraction of light is common in everyday life. Mirages or *apparently "wet" roads and waves of hot air rising* are formed because of the refraction of light. These images are created when light is bent upward due to the low index of refraction of hot air near the ground. Refraction is also responsible for the twinkling of stars. The air through which starlight travels continually varies because of temperature variation and turbulence. This causes the stars' images to shimmer and twinkle. Because of refraction a straight object appears to be bent when placed in water. Thus, a fish inside water is actually not in the place where it appears to be. The depth of an object placed in water appears to be less than its actual depth because of the refraction of light. Because of atmospheric refraction of light, the sun (or the moon) is seen for a short time before it actually rises above, or after it sets below, the horizon. This happens because the denser air near Earth refracts the light over the horizon. The sun on the horizon also appears flattered as a result of atmospheric refraction.

22.4 Total Internal Reflection and Fiber Optics

When light moves from one medium to a less optically dense medium (having less value for the index of refraction), the refracted ray bends away from the normal. The more the angle of incidence increases, the more the refracted ray diverges from the normal and at a certain angle of incidence the angle of refraction becomes 90°.

2007 Pearson Education, Inc., Upper Saddle River, NJ. All rights reserved. This material is protected under all copyright laws as they currently exist. No portion of this material may be reproduced, in any form or by any means, without permission in writing from the publisher.

This angle of incidence is called the **critical angle**, θ_c. From **Snell's law** [Equation (22.5)].

$$\sin \theta_c = \frac{n_2}{n_1} \text{ where } n_1 > n_2 \qquad (22.6)$$

If the second medium is air, $n_2 \approx 1$. Thus,

$$\sin \theta_c = \frac{1}{n}$$

When the angle of incidence is greater than the critical angle, all the light is reflected. This is called **total internal reflection**. Total internal reflection occurs for incidence angles outside a cone of apex angle $2\theta_c$.

Internal reflection gives rise to a diamond's brilliance when placed in air. Because the critical angle of glass is less than 45°, prisms with 45° and 90° angles are used to reflect light 90° and 180°. Total internal reflection by prisms is used in binoculars, periscopes, and other optical instruments.

Fiber Optics
Optical fibers are used to transmit light from one place to another, such as in optical communications and in endoscopy. Internal reflection forms the basis of **fiber optics**. As light travels through the fiber it undergoes many internal reflections until it emerges from the opposite end of the fiber. An optical fiber is composed of a core with a high index of refraction and a cladding with a low index of refraction.

Internal reflection is an exceptionally efficient process. Optical fibers can be used to transmit light over long distances with losses of only about 25% per kilometer. Fibers whose diameters are about 10 μm are grouped

© 2007 Pearson Education, Inc., Upper Saddle River, NJ. All rights reserved. This material is protected under all copyright laws as they currently exist. No portion of this material may be reproduced, in any form or by any means, without permission in writing from the publisher.

together in flexible bundles of 4- to 10-mm diameters. A fiber bundle of cross-sectional area of 1 cm^2 may contain as many as 50 000 individual fibers. Optical fibers are used in medical diagnostics such with an *endoscope* and a *cardioscope*.

22.5 Dispersion

A light is called *monochromatic* if it has one frequency (that is, one wavelength). A *white light* contains all component frequencies or colors. When a white light is refracted by a prism, it spreads out into a spectrum of colors. This phenomenon is called the **dispersion** of light.

The speed of light, and hence the index of refraction, depends on the frequency (that is, the wavelength) of the light. In general, the greater the frequency, the higher the index of refraction. As a result, different colors refract by different amounts, and after refraction white light is separated into its component colors.

Diamond is more dispersive than glass. In addition to its brilliance because of internal reflections, a cut diamond shows a play of colors because of dispersion of the internally reflected light. Rainbows are caused by the dispersion of sunlight in rain droplets.

269

007 Pearson Education, Inc., Upper Saddle River, NJ. All rights reserved. This material is protected under all copyright laws as they currently exist. No portion of this material may be reproduced, in any form or by any means, without permission in writing from the publisher.

Hints and Suggestions for Solving Problems

1. Use of the law of reflection ($\theta_i = \theta_r$) and Snell's law of refraction ($n_1 \sin \theta_1 = n_2 \sin \theta_2$) will be helpful for solving most problems in Sections 22.1–22.4.

2. A common mistake made when using the condition $\sin \theta_c = n_2/n_1$ to determine the critical angle or an unknown index of refraction is to use the wrong values for n_2 and n_1. Remember that the left-hand side is less than 1 (since the magnitude of the sine of an angle is always less than 1 except when the angle is 90°, in which case the value is 1). Thus, n_2 must be less than n_1 in this expression.

© 2007 Pearson Education, Inc., Upper Saddle River, NJ. All rights reserved. This material is protected under all copyright laws as they currently exist. No portion of this material may be reproduced, in any form or by any means, without permission in writing from the publisher.

CHAPTER 23
MIRRORS AND LENSES

This chapter discusses image formation by plane mirrors, spherical mirrors, and lenses. It also discusses mirror and lens equations, ray diagrams, spherical aberration, and chromatic aberration.

Key Terms and Concepts

Plane mirror
Real and virtual images
Magnification
Concave and convex mirrors
Focal point
Ray diagram
Spherical aberration
Converging and diverging lenses
Thin-lens equation
Lens combination
Chromatic aberration
Astigmatism
Lens-maker's equation
Lens power
Diopter

23.1 Plane Mirrors

Mirrors are smooth reflecting surfaces, usually made of polished metal or glass. If a mirror is perfectly flat, it is called a **plane mirror**.

A mirror forms an image of an object by reflection of light. For a plane mirror, the image appears to form behind or "inside" the mirror; this is because after reflection the rays

©2007 Pearson Education, Inc., Upper Saddle River, NJ. All rights reserved. This material is protected under all copyright laws as they currently exist. No portion of this material may be reproduced, in any form or by any means, without permission in writing from the publisher.

appear to the eye to originate from behind the mirror. Such an image is called a **virtual image**, as shown. On the other hand, a spherical mirror can produce images where the light rays actually meet to form the image. This type of image is called a **real image**.

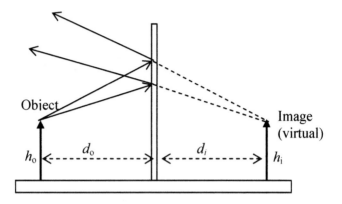

The law of reflection is always followed in the formation of the image. Images formed by plane mirrors have the following characteristics:

(i)   The image is upright and virtual.

(ii)   The distance (d_o) of the object (in front of the mirror) = the distance (d_i) of the image (behind the mirror).

(iii)   The image is the same size as the object. The image shows right-left reversal relative to the object.

In general, the magnification of an image is defined as

© 2007 Pearson Education, Inc., Upper Saddle River, NJ. All rights reserved. This material is protected under all copyright laws as they currently exist.
No portion of this material may be reproduced, in any form or by any means, without permission in writing from the publisher.

$$M = \frac{\text{image height}}{\text{object height}} = \frac{h_i}{h_o} \qquad (23.1)$$

$M = 1$, for the image formed by a plane mirror.

To see the image of a person's full body, the size of the mirror needs to be only half the height of the person.

23.2 Spherical Mirrors

A **spherical mirror** is a reflecting surface with spherical symmetry. If the inside surface (that is, the concave surface) of a section from a spherical surface is polished, it forms a **concave mirror**. If the outside surface (that is, the convex surface) is polished, it forms a **convex mirror**.

The **center of curvature** (C) is the center of the sphere of which the mirror is a section. The radial line passing through the center C is called the *optic axis*, and it meets the mirror surface at the *vertex*. The distance between the vertex and the center of curvature is the radius of the sphere, which is called the **radius of curvature** (R).

Rays of light incident parallel to the optic axis of a concave mirror, after reflection by the mirror, meet on a point, called the **focal point**. That is why a concave mirror is called a **converging mirror**. On the other hand, rays of light incident parallel to the optic axis of a convex mirror, after reflection by the mirror, diverge and appear to originate from its focal point. That is why a convex mirror is called a **diverging mirror**. The distance of the focal point from the vertex is called the **focal distance**, f. For a spherical mirror,

2007 Pearson Education, Inc., Upper Saddle River, NJ. All rights reserved. This material is protected under all copyright laws as they currently exist. No portion of this material may be reproduced, in any form or by any means, without permission in writing from the publisher.

$$f = \frac{1}{2} R \qquad\qquad (23.2)$$

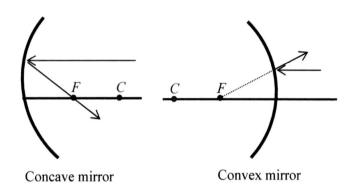

Concave mirror · · · · · · · · Convex mirror

Ray Diagrams

Location, size, and orientation of images formed by spherical mirrors can be obtained qualitatively using a **ray diagram.** The method involves drawing rays coming from one or more points of the object drawn carefully to the scale or using a relation called the **mirror equation**.

The law of reflection is applied and the following properties are used typically in a ray tracing:

1. A ray parallel to the optic axis is called a **parallel ray**, that after reflection passes through the focal point (for a concave mirror) or seems to originate at the focal point (for a convex mirror).

2. A ray passing through the center of curvature (for a concave mirror) or directed toward the center of curvature (for a convex mirror) is called a **chief ray** or **radial ray**, that after reflection follows the same path.

274

© 2007 Pearson Education, Inc., Upper Saddle River, NJ. All rights reserved. This material is protected under all copyright laws as they currently exist. No portion of this material may be reproduced, in any form or by any means, without permission in writing from the publisher.

3.   A ray passing through the focal point (for a concave
     mirror) or directed toward the focal point (for a convex
     mirror) is called a **focal ray**, that after reflection
     becomes parallel to the optic axis.

It is customary to use the tip of the object as the origin of
these rays for a ray diagram. Since the reflected rays at
any distance from a convex mirror diverge, a convex
mirror always forms a virtual image. For a concave
mirror, if the object distance is more than the focal length
of the mirror, the image is real and inverted. If the object
is placed within the focal length, the image is virtual,
upright and enlarged. The image is at infinity, if the object
is placed at the focal point.

The mathematical relation between the object distance, d_o,
and the image distance, d_i, of a mirror is known as the
spherical mirror equation. The mirror equation for a
mirror of focal length f is

$$\frac{1}{d_o} + \frac{1}{d_i} = \frac{1}{f} = \frac{2}{R} \qquad (23.3)$$

The **magnification factor**, M, is the ratio of image height,
h_i, to object height, h_o. The magnification is also given by

$$M = -\frac{d_i}{d_o} \qquad (23.4)$$

$$M = \frac{h_i}{h_o} = -\frac{d_i}{d_o}$$

275

©2007 Pearson Education, Inc., Upper Saddle River, NJ. All rights reserved. This material is protected under all copyright laws as they currently exist.
No portion of this material may be reproduced, in any form or by any means, without permission in writing from the publisher.

Sign Convention for Spherical Mirrors
Focal length
+ for a concave mirror ($+f$ or $+R$)
– for a convex mirror ($-f$ or $-R$)
Object distance and image distance
+ if the object or image is in front of the mirror ($+d_o$ or $+d_i$)
– if the object or image is behind the mirror ($-d_o$ or $-d_i$)
Image orientation (magnification)
+ if the image is upright with respect to the object ($+M$)
– if the image is inverted with respect to the object ($-M$)

Spherical Mirror Aberrations

According to the small-angle approximation, incident rays of light parallel to and near the axis of a concave spherical mirror converge at the focal point of the mirror. However, when the incident parallel rays are not near the optic axis, after reflection, these rays converge in front of the focal point. This effect gives rise to a blurred image of the object. This defect of the image formation is known as **spherical aberration**.

In a parabolic mirror, all the incident rays converge at the focal point, and thus spherical aberration is absent in the image. That is why parabolic mirrors are used in astronomical telescopes.

© 2007 Pearson Education, Inc., Upper Saddle River, NJ. All rights reserved. This material is protected under all copyright laws as they currently exist.
No portion of this material may be reproduced, in any form or by any means, without permission in writing from the publisher.

23.3 Lenses

A **lens** is a transparent piece of a material that can form an image by refraction. When rays of light pass through a lens, they are deviated from their original paths because of refraction. Biconvex spherical lenses have both surfaces convex, and biconcave spherical lenses have both surfaces concave.

A biconvex lens is called a **converging lens**, since it converges incident parallel rays of light (such as the light from the sun) to its focal point. Refraction by a biconvex lens can be analyzed by approximating the lens with two prisms placed base to base.

A concave lens is called a **diverging lens**, since it diverges incident parallel rays of light. When the refracted rays are extended back, they appear to originate at the focal point of the concave lens.

Lens shapes vary widely and lenses are categorized as converging or diverging. In general, a converging lens is thicker at its center, and a diverging lens is thinner at its center than at the periphery.

As with mirrors, ray tracing (using parallel rays, chief rays, and focal rays) is a convenient way to study images formed by a lens.

1. A ray parallel to the lens axis is called a **parallel ray** (1 in the figure), which after refraction passes through the focal point (for a converging lens) or seems to originate at the focal point (for a diverging lens).

©2007 Pearson Education, Inc., Upper Saddle River, NJ. All rights reserved. This material is protected under all copyright laws as they currently exist.
No portion of this material may be reproduced, in any form or by any means, without permission in writing from the publisher.

2. A ray passing through the center of a lens is called a **chief ray** or **central ray** (2 in the figure), and goes undeviated.

3. A ray passing through the focal point (for a converging lens) or directed toward the focal point (for a diverging lens) is called a **focal ray** (3 in the figure), which after refraction, becomes parallel to the lens axis.

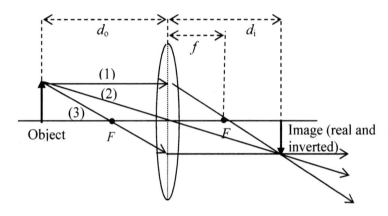

The mathematical relation between the object distance, d_o, and the image distance, d_i, of a thin lens is called the **thin-lens equation**. The equation for a lens of focal length f is

$$\frac{1}{d_o} + \frac{1}{d_i} = \frac{1}{f} \tag{23.5}$$

The **magnification factor** is given by

$$M = -\frac{d_i}{d_o} \tag{23.6}$$

© 2007 Pearson Education, Inc., Upper Saddle River, NJ. All rights reserved. This material is protected under all copyright laws as they currently exist. No portion of this material may be reproduced, in any form or by any means, without permission in writing from the publisher.

Sign Convention for Thin Lenses
Focal length
+ for a converging lens $(+f)$

− for a diverging lens $(-f)$

Object distance
+ if the object is in front of the lens $(+d_o)$

− if the object is behind the lens $(-d_o)$

Image distance
+ if the image is on the opposite of the lens from the object $(+d_i)$

− if the image is on the same side of the lens as the object $(-d_i)$

Image orientation (magnification)
+ if the image is upright with respect to the object $(+M)$

− if the image is inverted with respect to the object $(-M)$

Combination of Lenses

Lenses in combination are used in optical instruments including microscopes, telescopes, and cameras.

For a system containing two lenses in combination:

(i)   The image produced by the first lens acts as the *object* for the second lens,

(ii)   The total magnification produced by the combination is equal to the *product* of the magnifications produced by each lens individually. That is,

$$M_{total} = M_1 M_2 \qquad (23.7)$$

007 Pearson Education, Inc., Upper Saddle River, NJ. All rights reserved. This material is protected under all copyright laws as they currently exist. No portion of this material may be reproduced, in any form or by any means, without permission in writing from the publisher.

The sign conventions for M_1 and M_2 carry through the product to indicate from the sign of M_{total} whether the final image is upright or inverted.

23.4 The Lens Maker's Equation

The refraction by a lens depends on the shapes of the surfaces of the lens and on the index of refraction. The equation for a thin lens placed in air that relates to the index of refraction and the radii of curvature of the two surfaces (R_1 and R_2) of the lens is called the **lens maker's equation** and is given by

$$\frac{1}{f} = (n-1)\left(\frac{1}{R_1} - \frac{1}{R_2}\right) \qquad (23.8)$$

According to the sign convention, the radius of curvature (R) is positive when the center of curvature (C) is on the side of the lens from which light emerges. Focal length (f) is positive for a converging lens.

Lens Power: Diopters
The power of a lens gives the ability of a lens to refract light. The lens power is defined as

$$\text{Power} = \frac{1}{f} \qquad (23.9)$$

The SI unit of lens power is called the **diopter** if f is measured in meters.

$$1 \text{ diopter (D)} = 1 \text{ m}^{-1}$$

Power is positive for a converging lens and negative for a diverging lens. Optometrists characterize lenses in terms of diopters. A power of +2 D means a converging lens of focal length 0.5 m or 50 cm.

© 2007 Pearson Education, Inc., Upper Saddle River, NJ. All rights reserved. This material is protected under all copyright laws as they currently exist. No portion of this material may be reproduced, in any form or by any means, without permission in writing from the publisher.

*23.5 Lens Aberrations

Spherical Aberration

In a converging lens the incident parallel rays after refraction do not come together to a common focus, causing a blurred image. This defect is **spherical aberration**. The rays close to the axis meet at a point farther from the lens than the rays passing through the periphery.

Spherical aberration is reduced by using an aperture (to reduce the effective area of the lens) or using a lens combination (so that the aberration by one lens is compensated for by the other lens).

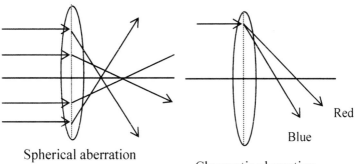

Spherical aberration

Chromatic aberration

Chromatic Aberration

Chromatic aberration arises because different colors are focused at different points on the axis of the lens. This type of aberration arises because the index of refraction of the material from which the lens is made is different for different colors. As a result, the lens forms a fringe of colors around the image.

©2007 Pearson Education, Inc., Upper Saddle River, NJ. All rights reserved. This material is protected under all copyright laws as they currently exist. No portion of this material may be reproduced, in any form or by any means, without permission in writing from the publisher.

Chromatic aberration can be prevented or reduced by using a compound lens system. A suitable combination of converging and diverging lenses can make the effects of the divergent refraction cancel and bring different colors to the same focus. Such a lens combination is said to be an *achromatic doublet.*

Astigmatism

When a circular cone of light from an off axis source falls on the surface of a convex lens some distance away, the light forms an elliptic area on the lens, and the rays entering the major and minor axes of the ellipse focus at different points after passing through the lens. This is called **astigmatism.**

Hints and Suggestions for Solving Problems

1. An image formed by a plane mirror can easily be drawn using the law of reflection (angle of incidence = angle of reflection). Keep in mind that the image is virtual, has the same size as the object, and is formed on the other side of the mirror at a distance that is the same as the distance of the object from the mirror.

2. When using the expression $\dfrac{1}{d_o} + \dfrac{1}{d_i} = \dfrac{1}{f}$, a common mistake that students make is to determine the value of the left-hand side and use that as the result for f. Remember that the left-hand side gives the value for $1/f$, so you need to invert the result to find f. Also, remember that you cannot add d_o and d_i and take the

© 2007 Pearson Education, Inc., Upper Saddle River, NJ. All rights reserved. This material is protected under all copyright laws as they currently exist. No portion of this material may be reproduced, in any form or by any means, without permission in writing from the publisher.

reciprocal of the sum to get the value of the left-hand side.

3. Keep in mind the sign conventions for mirrors. The focal length for a concave mirror is positive, whereas the focal length for a convex mirror is negative. Object and image distances are positive in front of the mirror and negative if behind the mirror.

4. The magnification is positive if the image is upright, and negative if the image is inverted. Suggestions 1 through 4 will be useful for solving problems in Sections 23.1 and 23.2.

5. Remember that the focal length of a convex lens is considered to be positive, and the focal length of a concave lens is considered to be negative. The sign conventions for object and image distances as well as for magnification for image formation by lenses are the same as those for images formed by mirrors. Keep these in mind when solving problems in Section 23.3.

6. It is a good habit to draw the ray diagram when solving a problem. If you draw the sketch to scale, your result from the sketch should agree with the numerical solution of the problem.

7. For the power of a lens in diopters, the focal distance must be in meters. Note this, since in many situations focal length of a lens is reported in centimeters or millimeters. The power of a lens is positive if it is a converging lens, but negative if it is a diverging lens.

2007 Pearson Education, Inc., Upper Saddle River, NJ. All rights reserved. This material is protected under all copyright laws as they currently exist. No portion of this material may be reproduced, in any form or by any means, without permission in writing from the publisher.

CHAPTER 24
PHYSICAL OPTICS: THE WAVE
NATURE OF LIGHT

This chapter discusses the wave properties of light,
including interference, diffraction, and polarization. It also
discusses some applications of these wave properties and
atmospheric scattering of light.

Key Terms and Concepts

> Constructive and destructive interference
> Coherent sources
> Thin-film interference
> Newton's rings
> Non-reflecting coating
> Diffraction
> Single-slit diffraction
> Diffraction grating
> X-ray diffraction
> Bragg's law
> Polarization
> Polarizing angle or Brewster angle
> Birefringence
> Dichroism
> Malus's law
> Scattering
> Rayleigh scattering

Geometrical optics ignores the wave properties of light. It
can explain image formations by reflection and refraction
of light but cannot explain important phenomena of light
such as interference, diffraction, and polarization.
Physical optics or **wave optics** takes into account the

© 2007 Pearson Education, Inc., Upper Saddle River, NJ. All rights reserved. This material is protected under all copyright laws as they currently exist.
No portion of this material may be reproduced, in any form or by any means, without permission in writing from the publisher.

wave properties of light and is needed to explain these phenomena.

24.1 Young's Double-Slit Experiment

The wave nature of light that gives rise to the interference effects can be seen in **Young's double-slit experiment**. The interference of light is not easily observed because of the short wavelength (about 10^{-7} m) of light. Also, to produce a steady interference pattern the interfering sources must be *coherent*, that is, the sources must have a constant phase relationship to one another.

In the experiment a monochromatic beam of light, after passing through a small slit, illuminates a double slit on a screen. Although the original light source is incoherent, the double slit acts as two coherent light sources, since any random change in the original source will simultaneously occur at these light sources, thus keeping the phase difference between these sources always constant. The light sources from the double slit interfere to form bright and dark interference "fringes" on the distant screen, showing constructive interference and destructive interference of light at different points.

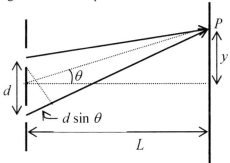

2007 Pearson Education, Inc., Upper Saddle River, NJ. All rights reserved. This material is protected under all copyright laws as they currently exist. No portion of this material may be reproduced, in any form or by any means, without permission in writing from the publisher.

From the geometry of the experiment, if d is the distance between the two slits and P be a point on the screen where lights from the slits interfere, the path-length difference between the rays that meet at P is $\Delta L = d \sin \theta$.

Since for constructive interference the path difference between two sources is an integral multiple of the wavelength, λ,

$$\Delta L = n\lambda \qquad \text{for } n = 0, 1, 2,... \qquad (24.1)$$

Similarly for destructive interference at P

$$\Delta L = m\lambda/2, \qquad \text{for } m = 1, 3, 5,... \qquad (24.2)$$

Thus, bright fringes (constructive interference) occur at different angles θ given by

$$d \sin \theta = n\lambda, \qquad \text{for } n = 0, 1, 2,... \qquad (24.3)$$

Different values of n correspond to different bright fringes. n is called the *order number* of the bright fringe.

The location of a given bright fringe in terms of its linear distance, y, from the central bright fringe is given by
$$y = L \tan \theta = L \sin \theta \text{ (for } \theta \text{ small)}$$
From the preceding equation and Equation (24.3)

$$y_n \approx \frac{nL\lambda}{d} \qquad (24.4)$$

where y_n is the location of the n-th bright fringe from the central maximum on either side. Since y is proportional to λ, for a white light the central fringe ($n = 0$) is white but other orders contain a spectrum of colors with red farther out. The preceding equation is used to determine the wavelength of light from this experiment.

286

© 2007 Pearson Education, Inc., Upper Saddle River, NJ. All rights reserved. This material is protected under all copyright laws as they currently exist. No portion of this material may be reproduced, in any form or by any means, without permission in writing from the publisher.

24.2 Thin-Film Interference

Reflected light waves from objects can interfere with one another. Thin-film interference is a result of interference of light reflected from the opposite surfaces of a film.
A light wave may change its phase because of reflection.

- The phase of a light wave changes by 180° when it is reflected from a region with a higher index of refraction.
- The phase does not change when light is reflected from a region with a lower index of refraction.

The index of oil is more than that of air and water. For an oil film on water, thus, there is a 180° phase shift for light reflected from the air-oil surface but there is no phase shift of the reflected light at the oil-water surface.

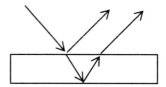

If the path length of the wave in the oil film is an odd multiple of the wavelength, the reflected rays would be in phase (as a result of the 180° phase shift of the incident wave) and would interfere constructively. Light of this color would be reflected. Thus, $2t = \lambda'/2$, where λ' is the wavelength in the oil; $\lambda' = \lambda/n$, where λ is the wavelength in air and n is the index of refraction of oil.

$$2t = m\lambda' \text{ (constructive interference)} \qquad (24.5)$$

$$2t = m\lambda'/2 \text{ (destructive interference)} \qquad (24.6)$$

287

©2007 Pearson Education, Inc., Upper Saddle River, NJ. All rights reserved. This material is protected under all copyright laws as they currently exist.
No portion of this material may be reproduced, in any form or by any means, without permission in writing from the publisher.

Since oil and soap films generally have different thickness in different places, different colors interfere constructively in different places, resulting in a vivid display of colors.

Thin film interference is seen by air film between two glass slides (such as two microscopic slides). The bright color of a peacock arises because of interference between the layers of fibers in its feathers. Light reflected from successive layers interferes constructively, giving bright colors.

Note that destructive interference does not mean that light energy is destroyed, it means that light energy is not present in that particular location. Analysis shows that light energy is redistributed and appears only at the location of bright fringes.

An application of thin-film interference is seen in non-reflecting coatings for lenses. Glass has a greater index of refraction than the non-reflecting film. Thus, the phase shifts of incident light occur at both the air-film and film-glass interfaces. In such a case, the condition for destructive interference is $2t = \lambda'/2$. Since $\lambda' = \lambda/n_1$, this gives the minimum film thickness as

$$t = \frac{\lambda}{4\,n_1} \tag{24.7}$$

where n_1 is the index of refraction of the film.

Optical Flats and Newton's Rings
Optical flats are made by grinding and polishing glass plates until they are as flat as desired. The degree of flatness is checked by forming a thin air wedge between the plate and a surface. If the surface is smooth, a regular and symmetric interference pattern is observed.

288

© 2007 Pearson Education, Inc., Upper Saddle River, NJ. All rights reserved. This material is protected under all copyright laws as they currently exist. No portion of this material may be reproduced, in any form or by any means, without permission in writing from the publisher.

If a curved piece of glass with a spherical cross section is placed on a flat glass, the interference pattern formed by reflection consists of concentric circular rings of alternate darkness and brightness. These are called *Newton's rings*.

The center of the rings, where the glass plates meet, is a dark spot. Fringes are more closely spaced as their distance from the center increases, since the curved surface of the upper glass moves away from the lower glass plate at a progressively faster rate.

24.3 Diffraction

Bending of light as it passes through a small hole or around an obstacle is called **diffraction**. Diffraction is a property of all waves. Diffraction of sound is quite evident since someone can talk from another room and we can easily hear. Diffraction of light is very small compared with that of sound, since its wavelength is very short (about 10^{-7} m for light compared with about a meter for sound). In general, *the longer the wavelength compared to the width of the opening, the greater the diffraction.*

The diffraction of light can be observed if a beam of monochromatic light passes through a single narrow slit.

According to Huygens' principle, each point on the slit behaves as a source of a secondary wave that radiates energy. The interference between these secondary waves gives rise to the diffraction pattern.

It can be shown that a dark fringe or *destructive interference* occurs at an angle θ in single-slit diffraction for

2007 Pearson Education, Inc., Upper Saddle River, NJ. All rights reserved. This material is protected under all copyright laws as they currently exist. No portion of this material may be reproduced, in any form or by any means, without permission in writing from the publisher.

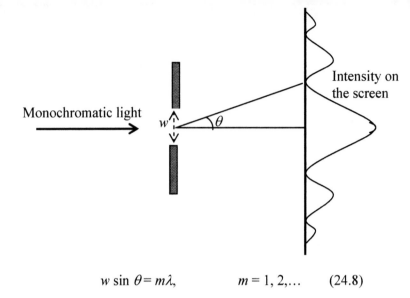

$$w \sin \theta = m\lambda, \qquad m = 1, 2,... \qquad (24.8)$$

In the preceding expression w is the width of the slit and λ is the wavelength of the light. Bright fringes appear halfway between successive dark fringes. The central fringe is also bright and is about twice as wide as the other bright fringes.

Using small-angle approximation, it can be shown that the distances of the dark fringes on either side of the central maximum is:

$$y_m = \frac{mL\lambda}{w} \qquad \text{for } m = 1, 2, 3, ... \qquad (24.9)$$

It is clearly evident from the Equation (24.9) that
(i) For a given slit width, the longer the wavelength, the wider the diffraction pattern, and
(ii) For a given wavelength, the narrower the slit width, the wider the diffraction pattern.

© 2007 Pearson Education, Inc., Upper Saddle River, NJ. All rights reserved. This material is protected under all copyright laws as they currently exist.
No portion of this material may be reproduced, in any form or by any means, without permission in writing from the publisher.

The width of fringes is given by

$$2y_1 \text{ (central maximum)} = \frac{2L\lambda}{w} \qquad (24.10)$$

$$y_{m-1} - y_m = \frac{L\lambda}{w} \qquad (24.11)$$

Diffraction Gratings

A large number of parallel, closely spaced slits is called a **diffraction grating**. The diffraction pattern consists of sharp and widely spaced bright fringes. If light is transmitted through a grating, it is called a *transmission grating*. If the light is reflected, it is called a *reflection grating*. The narrow grooves of a compact disk act as a reflection diffraction grating, producing colorful displays.

Bright fringes (called interference maxima) are created by constructive interference between light waves from different slits in the grating. The locations of the interference maxima are given by

$$d \sin \theta = n\lambda \qquad \text{for } n = 0, 1, 2, 3, \dots \qquad (24.12)$$

where d is the separation between two consecutive slits, θ is the angle of diffraction, and n is the order of interference.

Precision gratings may have 30 000 lines per centimeter or more. Diffraction gratings are widely used in spectroscopy to produce a spectrum and for determination of wavelength from the geometrical measurements of an experiment. A diffraction grating is often characterized by the number of lines, N, per centimeter. $d = 1/N$ if d is in cm.

©2007 Pearson Education, Inc., Upper Saddle River, NJ. All rights reserved. This material is protected under all copyright laws as they currently exist. No portion of this material may be reproduced, in any form or by any means, without permission in writing from the publisher.

Diffraction gratings are used in *spectrometers*. In a spectrometer, the materials are illuminated with lights of different wavelengths. The grating is rotated so that the sample is illuminated with a succession of different wavelengths.

Since $\sin \theta$ cannot exceed unity, it is seen from Equation (24.12) that the order number that can be observed is limited by

$$\frac{n\lambda}{d} \leq 1 \quad \text{or} \quad n_{max} \leq \frac{d}{\lambda}$$

X-ray Diffraction

A diffraction grating with appropriate spacing can be used to determine the wavelength of a wave. The regular spacing of the atoms in a crystalline solid act as a diffraction grating for incident X-rays to produce a diffraction pattern since the spacing in crystals is about 10^{-8} cm, the same as the order of wavelength of X-rays. It can be shown that the condition for constructive interference is

$$2d \sin \theta = n\lambda \text{ for } n = 1, 2, 3, \dots \qquad (24.13)$$

where d is the distance between the crystal's internal planes. This relation is called **Bragg's law**.

Because of their short wavelength, X-rays can be used as a diffraction probe to study the internal structure of proteins, DNA, and other molecules.

24.4 Polarization

The **polarization** of light refers to the direction of its oscillating electric field. Ordinary light is *unpolarized*. It is a combination of many waves with polarization in

292

© 2007 Pearson Education, Inc., Upper Saddle River, NJ. All rights reserved. This material is protected under all copyright laws as they currently exist. No portion of this material may be reproduced, in any form or by any means, without permission in writing from the publisher.

random directions. If there is some preferential orientation of the electric vectors, the light is said to be partially polarized. If the field vectors oscillate in only one plane, the light is *plane polarized*, or *linearly polarized*. Polarization of light is evidence that light is a transverse wave.

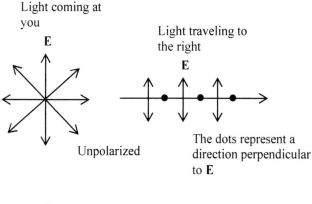

Light coming at you

Unpolarized

Light traveling to the right

The dots represent a direction perpendicular to **E**

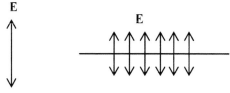

Linearly polarized

Polarization by Selective Absorption (Dichroism)

Some crystals (such as tourmaline) absorb one of the polarized components of incident light more than the other. This is called **dichroism**.

Materials (such as synthetic polymer materials) can transmit the components of the electric field along a preferred direction and absorb components perpendicular to this direction. The preferred direction is called the

293

2007 Pearson Education, Inc., Upper Saddle River, NJ. All rights reserved. This material is protected under all copyright laws as they currently exist. No portion of this material may be reproduced, in any form or by any means, without permission in writing from the publisher.

transmission axis or **polarization direction** of the polarizer. These materials are called polarizers.

If a light with intensity I_0 passes through a polarizer with its transmission axis at an angle θ relative to its polarization, the transmitted intensity, I, is given by

$$I = I_0 \cos^2 \theta \qquad (24.14)$$

This relation is known as *Malus's law*. The transmitted light is polarized in the direction of the polarizer.

Once polarized light has been produced using a polarizing material, it is possible to use a second polarizing material to change the intensity and direction of polarization of the polarized light. The second polarizing material is called an *analyzer*.

Polarizers with their transmission axes at right angles are called crossed polarizers ($\theta = 90^\circ$). The transmitted intensity through crossed polarizers is zero.

Polarization by Reflection
When light reflects from a smooth surface such as from the surface of a calm lake, it is partially reflected and partially transmitted. The reflected light may be completely polarized, partially polarized, or unpolarized, depending on the angle of incidence. The electric field components parallel to the surface are reflected more strongly, producing partial polarization of the reflected light. When the reflected and refracted lights are 90° apart, the reflected light is completely polarized and the refracted light is partially polarized. This angle of incidence is called the **polarizing angle** (θ_p) or the **Brewster angle**. It can be

© 2007 Pearson Education, Inc., Upper Saddle River, NJ. All rights reserved. This material is protected under all copyright laws as they currently exist. No portion of this material may be reproduced, in any form or by any means, without permission in writing from the publisher.

shown that this angle is related to the indices of refraction of the two media by

$$\tan \theta_p = n_2/n_1 \text{ and } \theta_p = \tan^{-1}(n_2/n_1) \qquad (24.15).$$

where n_1 is the index of refraction of the first medium.

To reduce the glare of the reflected light, Polaroid sunglasses with a vertical transmission axis are used, which can block light polarized in the horizontal direction. Light from a rainbow is partially polarized since the reflection angle inside a water droplet is close to the Brewster angle.

Polarization by Double Refraction (Birefringence)
The speed of light is different in different directions for some *anisotropic* crystals such as quartz and calcite. Such materials exhibit **birefringence** or are double refracting. When a beam of unpolarized light propagates at an angle to a particular crystal axis, the beam is doubly refracted and separated into two components, as shown. The components are called the *ordinary ray* (o) and the *extraordinary ray* (e) that are linearly polarized in mutually perpendicular directions.

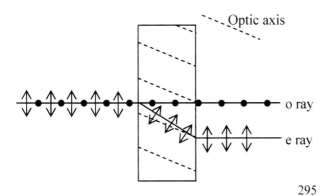

295

2007 Pearson Education, Inc., Upper Saddle River, NJ. All rights reserved. This material is protected under all copyright laws as they currently exist. No portion of this material may be reproduced, in any form or by any means, without permission in writing from the publisher.

There are organic compounds (such as sugar and tartaric acid) that can rotate the polarization direction of an incident beam of light. This property is called **optical activity**. The amount of rotation depends on the concentration of the active molecules in the medium.

*24.5 Atmospheric Scattering of Light

When light is incident on air molecules or any suspension of particles, some of the light is absorbed and reemitted. This phenomenon is called *scattering*. The scattering of sunlight causes polarization of scattered sky light, the blueness of the sky, and the redness of sunrises and sunsets.

Light scattered by atoms or molecules in the atmosphere is polarized when viewed at right angles to the Sun. When viewed at an intermediate angle, the light has an intermediate amount of polarization. Since the scattering of light with the greatest degree of polarization is at a right angle to the direction of the sun, at sunrise and sunset the scattered light from directly overhead has the greatest degree of polarization.

Why the Sky Is Blue
It has been found that the scattering, S, is inversely proportional to the fourth power of the wavelength of light, that is, $S \propto 1/\lambda^4$. This relationship is known as **Rayleigh scattering**. Thus, the scattering of sunlight by air molecules is most effective for light of short wavelength, such as blue light. This is why the sky appears blue.

Why Sunsets and Sunrises Are Red
When the sun is near the horizon (during sunrise or sunset), the sunlight travels a greater distance through the

296

© 2007 Pearson Education, Inc., Upper Saddle River, NJ. All rights reserved. This material is protected under all copyright laws as they currently exist. No portion of this material may be reproduced, in any form or by any means, without permission in writing from the publisher.

denser air near the Earth's surface and undergoes a great deal of scattering. The sunrise and sunset appear red when the sun is viewed through the atmosphere because most of the blue light has been scattered off in different directions by the air molecules (and by small foreign molecules). Red sunrises and sunsets become visible when there is a high-pressure air mass to the east and high-pressure air mass to the west, respectively. The sky of Mars is red because its atmosphere contains 95% carbon dioxide that preferentially scatters red wavelengths.

Hints and Suggestions for Solving Problems

1. For *constructive interference*, the path difference between the two waves is $\Delta L = n\lambda$, where $n = 0, 1, 2,\ldots$; and for *destructive interference*, the path difference is $\Delta L = m\lambda/2$, where $m = 11, 3, 5, \ldots$. This information will be useful for solving many problems in this chapter.

2. You may not need to remember the equations for the double-slit experiment; the conditions for bright and dark fringes can be easily derived from tip 1. The location of a fringe from the central fringe in Young's experiment is given by $y = L \tan \theta$. When you make small-angle approximations, $\tan \theta = \theta$, make sure θ is in radians.

3. To solve problems on interference in reflected beams remember that the phase of a light wave changes by 180° when it is reflected from a region with a higher index of refraction. Phase does not change when light

297

007 Pearson Education, Inc., Upper Saddle River, NJ. All rights reserved. This material is protected under all copyright laws as they currently exist. No portion of this material may be reproduced, in any form or by any means, without permission in writing from the publisher.

is reflected from a region with a lower index of refraction.

4. For problems of diffraction by a single slit, such as those in Section 24.3, use the equation $w \sin \theta = m\lambda$ where $m = 1, 2,....$ Remember that this equation provides the conditions for the dark fringes and *not* for the bright fringes.

5. The density of lines on a grating is usually provided as the number of lines per centimeter. Be sure to convert the density of lines to *number of lines per meter*. The equation $d \sin \theta = n\lambda$, $n = 0, 1, 2,...$ gives the locations of the *interference maxima*.

6. To determine the transmitted light intensity, I, such as problems in Section 24.4, using the law of Malus, $I = I_0 \cos^2 \theta$, note that θ is the angle with the polarization axis, not with the x-axis.

© 2007 Pearson Education, Inc., Upper Saddle River, NJ. All rights reserved. This material is protected under all copyright laws as they currently exist. No portion of this material may be reproduced, in any form or by any means, without permission in writing from the publisher.

CHAPTER 25
VISION AND OPTICAL INSTRUMENTS

This chapter focuses on common optical instruments. In particular, it describes human eye and vision defects, magnifying glasses, compound microscopes, and telescopes.

Key Terms and Concepts

    The human eye
    Near point and far point
    Nearsightedness and farsightedness
    Astigmatism
    The magnifying glass
    Angular magnification
    The compound microscope
    Reflecting telescope
    Astronomical telescope
    Terrestrial telescope
    Resolution
    Raleigh criterion
    Primary colors
    Complementary colors

25.1 The Human Eye

The optical function of the human eye is similar to that of a camera. A camera consists of a converging lens that focuses images of an object on a light-sensitive film. The eye also has a converging lens that focuses images on the light-sensitive lining on the rear surface inside the eyeball.

The eyeball is nearly spherical of diameter about 1.5 cm and filled with a jelly-like substance called the *vitreous*

299

007 Pearson Education, Inc., Upper Saddle River, NJ. All rights reserved. This material is protected under all copyright laws as they currently exist. No portion of this material may be reproduced, in any form or by any means, without permission in writing from the publisher.

humor. Light enters the human eye through the *cornea*, a clear fluid called *aqueous humor*, and a *crystalline lens*. The eye forms a real inverted image behind the eye on the **retina** by refraction. The eye can control the amount of light reaching the retina by changing an opening, called the *pupil*, at the center of the iris. The retina is light sensitive and consists of millions of structures called **rods** and **cones**. When stimulated by light, these structures send electric impulses through the optic nerves to the brain.

Ciliary muscles change the shape, and hence focal length, of the eye lens to focus the image on the retina. This process is called *accommodation*. When the eye is viewing a distant object, the muscles are relaxed, but when the eye is viewing a nearby object, the muscles are tensed to give a greater curvature to the lens.

The *far point* is the greatest distance at which the normal eye can see objects clearly. The normal far point distance is infinity. The *near point* is the closest distance of an object that can be seen clearly. The near point gradually recedes with age. For an average young person the near point distance is about 25 cm.

Vision Defects

The most common vision defects are:

> Nearsightedness (myopia),
> Farsightedness (hyperopia), and
> Astigmatism

Nearsightedness (or *myopia*) is a defect of the human eye in which clear vision is restricted to nearby objects. The far point is not at infinity but is instead at a finite distance from the eye. It arises because the eyeball is too long or the curvature of the cornea is too great. This defect can be

300

© 2007 Pearson Education, Inc., Upper Saddle River, NJ. All rights reserved. This material is protected under all copyright laws as they currently exist. No portion of this material may be reproduced, in any form or by any means, without permission in writing from the publisher.

corrected by using a suitable divergence lens in front of the eye.

Farsightedness (or *hyperopia*) is a defect of the human eye in which clear vision is restricted to distant objects. The near point is much farther from the eye than the typical value of 25 cm. It arises because the eyeball is too short or because of insufficient curvature of the cornea. This defect can be corrected by using a suitable convergence lens in front of the eye.

Both nearsightedness and farsightedness can be simultaneously treated using bifocals. A current technique to correct nearsightedness involves the use of a laser. A flap of material on the cornea surface is cut and pulled up. The laser is used to shape the surface area of the cornea to have the image of a distance object fall on the retina.

Astigmatism is another defect of human vision that arises due to a refracting surface (cornea or lens) being nonspherical. As a result, the eye has different focal lengths in different planes and an image may be distinct in one direction and blurred in another. Nonspherical lenses such as plano-convex cylindrical lenses are used to correct astigmatism.

Visual acuity is a measure of how vision is affected by the object distance. The result is expressed as a fraction such as 20/20. The numerator is the distance at which the test eye sees a standard symbol and the denominator is the distance at which the letter is clearly seen by a *normal* eye.

©2007 Pearson Education, Inc., Upper Saddle River, NJ. All rights reserved. This material is protected under all copyright laws as they currently exist. No portion of this material may be reproduced, in any form or by any means, without permission in writing from the publisher.

25.2 Microscopes

Microscopes are used to magnify objects so that we can see some features that cannot be normally seen.

The Magnifying Glass (A Simple Microscope)
A **magnifying glass** is a simple convex lens that can make objects appear to be many times larger than their actual size. A magnifying glass works by moving the near point closer to the eye. Since the distance is reduced, the angular size is increased.

The factor by which an image is enlarged is called the **angular magnification**, m, which is defined as the ratio of the angular size of the object viewed through the magnifying glass (θ) to the angular size of the object viewed without the magnifying glass (θ_o):

$$m = \frac{\theta}{\theta_o} \qquad (25.1)$$

The maximum angular magnification occurs when the image seen through the glass is at the eye's near point, $d_i = -25$ cm. It can be shown from the thin lens equation that the corresponding object distance is:

$$d_o = \frac{(25\,\text{cm})f}{25\,\text{cm} + f} \qquad (25.2)$$

This gives,

$$m = 1 + \frac{25\,\text{cm}}{f} \quad \textit{angular magnification} \quad (25.3)$$

$$\textit{for image at near point}$$

302

© 2007 Pearson Education, Inc., Upper Saddle River, NJ. All rights reserved. This material is protected under all copyright laws as they currently exist. No portion of this material may be reproduced, in any form or by any means, without permission in writing from the publisher.

For the image to be at infinity, the object must be at the focal point of the lens. In this case,

$$\theta = \approx \frac{y_o}{f} , \text{ and the angular magnification is}$$

$$m = \frac{25 \text{ cm}}{f} \qquad \textit{image at infinity} \qquad (25.4)$$

The Compound Microscope

To provide greater angular magnification beyond what is possible with a magnifying glass, a **compound microscope** is used. This microscope consists of two converging lenses in combination—an **objective** and an **eyepiece** or **ocular**. The converging lens with a relatively short focal length ($f_o < 1$ cm) is called the objective. The object to be viewed is placed just outside the focal length of the objective. The real, inverted, and enlarged image formed by the objective serves as the object of the eyepiece. The eyepiece has a relatively longer focal length (f_e is a few cm) and is positioned so that the image formed by the objective falls just inside its focal point. The eyepiece thus behaves as a magnifier and gives additional magnification.

The total magnification is the product of magnifications by the two lenses and is given by

$$m_{total} = \frac{(25 \text{ cm})L}{f_o f_e} \qquad (25.5)$$

where L is the distance between the lenses.

A modern compound microscope has interchangeable eyepieces with magnifications from about 5x to over 100x.

303

2007 Pearson Education, Inc., Upper Saddle River, NJ. All rights reserved. This material is protected under all copyright laws as they currently exist. No portion of this material may be reproduced, in any form or by any means, without permission in writing from the publisher.

The maximum magnification with a compound microscope is about 2000x.

25.3 Telescopes

A telescope is an optical instrument that is used for seeing and magnifying distant objects. There are two types of telescopes: refracting and reflecting.

Refracting Telescope
A refracting telescope works on the same principle as a compound microscope. Like a compound microscope, it uses two converging lenses in combination.

The objective is a large converging lens that focuses the light from distant objects at its focal point (f_o). The eyepiece is movable and has a relatively short focal length (f_e). It magnifies the image formed by the objective. For relaxed viewing, the image of the objective is at the focal point of the eyepiece and the distance between the lenses is ($f_o + f_e$).

The total angular magnification of a telescope for the final image at infinity is

$$m = -\frac{f_o}{f_e} \qquad (25.6)$$

The minus sign indicates that the image is inverted. The objective of a telescope should be large since the objects to be viewed are usually very dim, and its focal length should be long for greater magnification. This kind of telescope in which the image is inverted is called an **astronomical telescope**.

© 2007 Pearson Education, Inc., Upper Saddle River, NJ. All rights reserved. This material is protected under all copyright laws as they currently exist. No portion of this material may be reproduced, in any form or by any means, without permission in writing from the publisher.

A telescope in which the final image is upright is called a **terrestrial telescope**. A Galilean telescope uses a diverging lens as an eyepiece to produce an upright and virtual image. Galilean telescopes have a very narrow field of view and limited magnification. A better way to produce an upright image is to use a converging "erecting" lens between the objective and eyepiece of an astronomical telescope. This addition increases the length of the telescope but the length can be shortened by using internally reflecting prisms.

Reflecting Telescope

A **reflecting telescope** uses a mirror instead of a converging lens for the objective. For distant stars and galaxies it is important to gather enough light to see the object. This requires a large objective lens for a refractive telescope and encounters other associated difficulties. Mirrors can be much thinner and lighter, and these have many advantages. A reflecting telescope usually uses a large, concave, front-surface parabolic mirror. It does not have spherical aberration or chromatic aberration. The largest telescopes are all reflectors.

25.4 Diffraction and Resolution

The ability to visually distinguish closely spaced objects is called the **resolution** of a system. Diffraction places a natural limit on the resolution of a visual system. According to the **Raleigh criterion** *two images are said to be just resolved when the central maximum of one image falls on the first minimum of the diffraction pattern of the other image.*

2007 Pearson Education, Inc., Upper Saddle River, NJ. All rights reserved. This material is protected under all copyright laws as they currently exist. No portion of this material may be reproduced, in any form or by any means, without permission in writing from the publisher.

For the first minimum ($m = 1$) for a single slit of width w, $w \sin \theta = \lambda$. Thus, the minimum angle of resolution for a single slit is approximately

$$\theta_{min} = \frac{\lambda}{w} \qquad (25.7)$$

The apertures of optical instruments (such as a telescope or microscope) are usually circular. A circular aperture produces a diffraction pattern consisting of concentric alternate dark and bright fringes. The condition for the minimum angular separation for the images of two objects to be just resolved is

$$\theta_{min} = \frac{1.22\lambda}{D} \qquad (25.8)$$

where D is the diameter of the aperture. At large distances, the angular separation of the two headlights of an approaching car is less than that given by Equation (25.8), and the lights appear to originate from a single source. As the car comes closer the angular separation exceeds the value in Equation (25.8), and the lights are resolved.

The minimum separation, s, between two points whose images can just be resolved is called the **resolving power** of a microscope. For a microscope,

$$s = f\,\theta_{min} = \frac{1.22\lambda f}{D} \qquad (25.9)$$

© 2007 Pearson Education, Inc., Upper Saddle River, NJ. All rights reserved. This material is protected under all copyright laws as they currently exist. No portion of this material may be reproduced, in any form or by any means, without permission in writing from the publisher.

*25.5 Color

Color Vision

The cone receptors in the retina of the human eye are sensitive to light of wavelengths between 400 nm and 700 nm. Color is perceived because of a physiological response to excitation by light of the cone receptors. The brain perceives different wavelengths of visible light as different colors. The human eye is most sensitive near 550 nm, which is in the yellow-green region.

There are three types of cones in the retina corresponding to the red, green, and blue regions of the visible spectrum. *Color blindness* occurs when any type of cone is missing. Red, blue, and green are called **additive primary colors** because when light beams of these colors are projected onto a white screen, mixtures of them produce other colors. This technique is called the **additive method of color production**. Also, a combination of many pairs of colors (such as blue and yellow) appears white. The colors of such pairs are called **complementary colors**.

Objects show a color when illuminated with white light because they either reflect or transmit the wavelength of that color. For example, a red apple apprears red because it strongly reflects the red portions of white light and absorbs other colors. A red filter appears red because color pigments in the glass absorb colors of white light other than red and it transmits red. A mixture of yellow and blue paints produces green. This is because the yellow and green pigments of the paints absorb most colors except the yellow and nearby regions and except blue and nearby regions, respectively. The intermediate green region is not strongly absorbed by either pigment and hence the mixture appears green. When the **subtractive primary pigments**

307

©2007 Pearson Education, Inc., Upper Saddle River, NJ. All rights reserved. This material is protected under all copyright laws as they currently exist. No portion of this material may be reproduced, in any form or by any means, without permission in writing from the publisher.

(cyan, magenta, and yellow) are mixed, different colors are produced by subtractive absorption; this is called the **subtractive method of color production**. .

Hints and Suggestions for Solving the Problems

1. Remember the sign conventions for object distance, image distance, focal length, and magnification.

2. Remember that for the correction of nearsightedness a diverging lens is needed. For the correction of farsightedness a converging lens is needed.

3. The focal length of a lens is usually given in centimeters. Remember to convert the focal length to meters. The refractive power will be in diopters only if the focal length is expressed in meters.

4. Keep in mind that the total magnification of an object by a lens combination (such as compound microscopes and telescopes) is the product of the magnifications by the individual lenses.

5. Also keep in mind that for a lens combination, the image formed by the first lens acts as the object for the second lens in the combination. Tips 4 and 5 will be useful for solving problems in Sections 25.2 and 25.3.

6. The minimum angle of resolution for a circular aperture is $\theta_{min} = 1.22\lambda/D$ whereas the minimum angle of resolution for a slit is $\theta_{min} = \lambda/w$. The resolving power of a micrsopce is $s = 1.22\lambda f/D$. The preceding equations will be useful to solve problems in Section 25.4.

308

© 2007 Pearson Education, Inc., Upper Saddle River, NJ. All rights reserved. This material is protected under all copyright laws as they currently exist. No portion of this material may be reproduced, in any form or by any means, without permission in writing from the publisher.

CHAPTER 26
RELATIVITY

This chapter describes the theory of relativity. The special theory of relativity gives the true interpretation of space and time as well as of mass and energy. The special theory is needed to understand the dynamics of objects moving at speeds comparable to the speed of light. The general theory of relativity provides a new interpretation of gravity.

Key Terms and Concepts

> Inertial reference frame
> Newtonian relativity
> The ether
> Interferometer
> Michelson–Morley experiment
> Postulates of special relativity
> The relativity of simultaneity
> Relativity of time
> Time dilation
> Proper time
> Muon decay
> Length contraction
> Proper length
> The twin paradox
> Relativistic kinetic energy
> Relativistic momentum
> Rest energy
> Relativistic energy
> Mass-energy equivalence
> The general theory of relativity
> The equivalence principle
> Gravitational lensing

309

2007 Pearson Education, Inc., Upper Saddle River, NJ. All rights reserved. This material is protected under all copyright laws as they currently exist. No portion of this material may be reproduced, in any form or by any means, without permission in writing from the publisher.

Black hole
Schwartzschild radius
Event horizon
Relativistic velocity addition

Newton's laws apply only to macroscopic objects moving at small speeds. For objects moving at speeds comparable to the speed of light the special theory is needed to understand the dynamics of their motion. The special theory provides correct results for all speeds from zero to the speed of light.

26.1 Classical Relativity and the Michelson–Morley Experiment

A physical principle or law should be same regardless of the observer's frame of reference. **An inertial frame of reference** is that in which Newton's first law of motion holds. That is, an isolated object in this frame is either at rest or moves with a constant velocity. A non-inertial frame is one in which an isolated object appears to accelerate. Any reference frame that moves with a constant velocity relative to an inertial frame is also inertial. There is no one inertial frame that is preferred over another for the laws of mechanics. Thus, the laws of mechanics are the same in all inertial reference frames. This is called the *principle of classical or Newtonian relativity.*

The Absolute Reference Frame: the Ether
Maxwell's theory predicted that light is an electromagnetic wave that travels with a speed of 3×10^8 m/s in a vacuum. It was assumed that this speed must be referenced to a particular frame called *luminiferous ether* or simply **ether**.

310

© 2007 Pearson Education, Inc., Upper Saddle River, NJ. All rights reserved. This material is protected under all copyright laws as they currently exist. No portion of this material may be reproduced, in any form or by any means, without permission in writing from the publisher.

If the ether had existed, then presumably a true absolute rest frame would be identified. This was the objective of the Michelson and Morley experiment.

If the ether permeated all space, the Earth would be surrounded by it and the orbiting Earth would experience an "ether wind." Thus, observers could measure the speed of light in directions parallel and perpendicular to the orbital velocity of the Earth, and because of the relative motion the speeds should be different. However, the experiment showed that the speed of light is always the same. Thus, the ether did not exist and there was no absolute frame of reference. The constancy of the speed of light changed our views about space and time.

26.2 The Postulates of Special Relativity and the Relativity of Simultaneity

Einstein's special theory is based on two postulates. The first postulate is an extension of the Newtonian principle of relativity. In Einstein's theory the **principle of relativity** applies all the laws of physics including those in electricity and magnetism.

Postulate I
The laws of physics are the same in all inertial frames of reference.

This postulate implies that an absolute reference frame does not exist. That is, all inertial frames are equivalent.

Postulate II
The speed of light in a vacuum is the same in all inertial frames of reference and is independent of the motion of the source or the observer.

311

2007 Pearson Education, Inc., Upper Saddle River, NJ. All rights reserved. This material is protected under all copyright laws as they currently exist. No portion of this material may be reproduced, in any form or by any means, without permission in writing from the publisher.

Not only Michelson and Morley's experiment but most recent and accurate experiments confirm the constancy of the speed of light.

The Relativity of Simultaneity

Two events are said to be simultaneous if they occur at the same time. Einstein found from the constancy of the speed of light that the concept of simultaneity is relative. That is, *if two events are simultaneous in a particular inertial reference frame, they may not be simultaneous as measured in a different inertial frame.* The concept of the relativity of simultaneity violates our everyday experience since ordinary objects move at relative speeds much smaller than the speed of light.

26.3 The Relativity of Length and Time: Time Dilation and Length Contraction

Time Dilation

Time is not absolute but is relative. *Moving clocks are measured to run more slowly than clocks at rest in the observer's own frame of reference.* This stretching of time is known as **time dilation**.

An event is a physical occurrence that happens at a specified location at a specified time. The **proper time,** Δt_0, is the time separating two events occurring at the same location. If two events are separated by a proper time Δt_0 occurring in a reference frame that moves with a speed, v, relative to an observer, the dilated time measured by the observer is given by

312

© 2007 Pearson Education, Inc., Upper Saddle River, NJ. All rights reserved. This material is protected under all copyright laws as they currently exist. No portion of this material may be reproduced, in any form or by any means, without permission in writing from the publisher.

$$\Delta t = \frac{\Delta t_0}{\sqrt{1 - v^2/c^2}} \qquad (26.1)$$

In this expression c is the speed of light in a vacuum: $c = 3 \times 10^8$ m/s. Putting

$$\gamma = \frac{1}{\sqrt{1 - v^2/c^2}} \qquad (26.2)$$

we get,

$$\Delta t = \gamma \Delta t_0 \qquad (26.3)$$

(i) If $v = 0$, $\gamma = 1$. Thus, $\Delta t = \Delta t_0$, as expected.

(ii) As $v \to c$, $\Delta t \to \infty$. In this limit, a clock will slow down to the point of stopping.

(iii) For $v = 0.60c$, $\gamma = 1.25$. Thus, when 20 minutes have elapsed on a moving clock, on a clock at rest relative to the observer 1.25×20 or 25 minutes have elapsed.

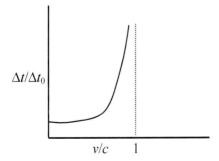

313

2007 Pearson Education, Inc., Upper Saddle River, NJ. All rights reserved. This material is protected under all copyright laws as they currently exist. No portion of this material may be reproduced, in any form or by any means, without permission in writing from the publisher.

Hafele and Keating experimentally verified time dilation in 1971. They carried an atomic clock in a jet plane and left an identical clock at rest in the laboratory. After flying the moving clock at a high speed, they found that the clock in the plane ran slower than the clock in the laboratory, in agreement with the special theory.

Time dilation applies equally to all physical processes, including biological processes and chemical reactions. An astronaut traveling in a spaceship ages more slowly than one who remains on Earth.

Muons are subatomic particles created in the atmosphere at a height of about 5000 m. When at rest, a muon exists for only 2.2×10^{-6} s before disintegrating. With such a short life, muons could never reach Earth's surface, even if they moved with the speed of light. The reason a large number of muons *do reach* Earth is that they age more slowly because of time dilation.

Length Contraction

Length is not absolute but is relative. The length of an object depends on its speed relative to an observer.

The **proper length**, L_0, is the distance (or length) between two points measured by an observer at rest with respect to them. An object of proper length L_0 moving with a speed v relative to an observer has a contracted length L as given by

$$L = L_0/\gamma = L_0 \sqrt{1 - \frac{v^2}{c^2}} \qquad (26.4)$$

Since γ is always greater than 1, the length has contracted by a factor of $1/\gamma$. If $v = c$, $L = 0$.

314

© 2007 Pearson Education, Inc., Upper Saddle River, NJ. All rights reserved. This material is protected under all copyright laws as they currently exist. No portion of this material may be reproduced, in any form or by any means, without permission in writing from the publisher.

Length contraction occurs only in the direction of motion. Lengths perpendicular to the direction of motion are not affected.

The Twin paradox

Time dilation and length contraction give rise to what is called the **twin paradox**. Consider two identical twins, one of whom goes on a high-speed journey to a distant star. When the traveling twin returns, he is younger than his Earth-bound twin.

The solution of this paradox lies in the fact that in leaving and returning to the Earth, the traveling twin has experienced accelerations and decelerations and thus was not always in an inertial reference frame. So the symmetry of special relativity does not apply.

According to the Earth-bound twin, the traveling twin's internal clock runs slower. So he returns home younger than the Earth-bound twin. According to the traveling twin, he has traveled a shorter distance because of the space contraction and thus he returns home younger than his Earth-bound twin.

26.4 Relativistic Kinetic Energy, Momentum, Total Energy, and Mass–Energy Equivalence

Relativistic Kinetic Energy

Einstein found that the *relativistic kinetic energy* of a particle of mass m moving at speed v is given by

2007 Pearson Education, Inc., Upper Saddle River, NJ. All rights reserved. This material is protected under all copyright laws as they currently exist. No portion of this material may be reproduced, in any form or by any means, without permission in writing from the publisher.

$$K = \left(\frac{1}{\sqrt{1 - v^2/c^2}} - 1 \right) mc^2 = (\gamma - 1)\, mc^2 \qquad (26.5)$$

When $v \ll c$, the preceding expression becomes $K = \frac{1}{2} mv^2$.

Relativistic Momentum

According to the special theory of relativity, the principle of momentum conservation holds if the expression for the momentum (called **relativistic momentum**) for an object moving with speed $\vec{v}$ is

$$\vec{p} = \frac{mv}{\sqrt{1 - \dfrac{v^2}{c^2}}} = \gamma m \vec{v} \qquad (26.6)$$

For small speeds, classical momentum and relativistic momentum are the same. As $v \to c$, $\vec{p} \to \infty$.

Relativistic Total Energy and Rest Energy: The Equivalence of Mass and Energy

According to relativity, the relativistic total energy of an object is

$$E = \frac{mc^2}{\sqrt{1 - \dfrac{v^2}{c^2}}} = \gamma mc^2 \qquad (26.7)$$

When the object is at rest ($v = 0$), its energy is called **rest energy**, E_0, which is given by

316

© 2007 Pearson Education, Inc., Upper Saddle River, NJ. All rights reserved. This material is protected under all copyright laws as they currently exist. No portion of this material may be reproduced, in any form or by any means, without permission in writing from the publisher.

$$E_0 = mc^2 \qquad (26.8)$$

The preceding expression means that mass and energy are equivalent. Energy has mass, and mass has energy.

The rest mass of an electron is 9.109×10^{-31} kg. This is equivalent to rest energy $E_0 = 0.511$ MeV.

From equation (26.7), the relativistic kinetic energy

$$K = (\gamma - 1)\, mc^2 = E - E_0.$$

In nuclear plants a small decrease in nuclear mass due to various nuclear reactions is converted to electrical energy. The sun also generates energy from the conversion of its mass to energy.

26.5 The General Theory of Relativity

The special theory deals with objects or frames moving at a constant velocity with respect to the others. The **general theory of relativity** deals with accelerating systems and it is essentially a new theory of gravitation.

The Principle of Equivalence
The general theory of relativity is based on the **principle of equivalence** according to which: *An inertial reference frame in a uniform gravitational field is equivalent to a reference frame in the absence of a gravitational field that has a constant acceleration with respect to that inertial frame.* This also means that *no experiment in a closed system can distinguish between the effects of a gravitational field and the effects of acceleration.*

317

2007 Pearson Education, Inc., Upper Saddle River, NJ. All rights reserved. This material is protected under all copyright laws as they currently exist. No portion of this material may be reproduced, in any form or by any means, without permission in writing from the publisher.

Light and Gravitation

The principle of equivalence of the general theory of relativity leads to important predictions including the bending of light in a gravitational field. The amount of bending depends on the strength of gravity. The general theory views a gravitational field as a warping of space and time.

Gravity causes a curvature in the space-time warp. As the light passes through the curved surface of space-time, it bends. The gravitational bending of light from a distant star as it passes close to the sun was verified during a solar eclipse in 1919.

Gravitational Lensing

Another effect of the gravitational bending of light is *gravitational lensing*. The bending of light by a massive object such as a galaxy or cluster of galaxies gives rise to multiple images of more distant objects such as quasars. Such effects have been observed recently by the Hubble Space Telescope images of distant quasars.

Black Holes

When a star collapses to a very small size, its gravitational field can become extremely intense. In some cases gravity can become so strong that it will bend light to a point that it cannot escape. Such a star does not emit any light of its own and is called a **black hole**. Evidence of black holes was found in the center of galaxies and in some constellations. An estimate of the size of a black hole can be determined from the escape speed, which from the surface of a spherical body of mass M and Radius R is given by

© 2007 Pearson Education, Inc., Upper Saddle River, NJ. All rights reserved. This material is protected under all copyright laws as they currently exist
No portion of this material may be reproduced, in any form or by any means, without permission in writing from the publisher.

$$v_{esc} = \sqrt{\frac{2GM}{R}} \qquad (26.9)$$

In order for an astronomical object of mass M to be a black hole, v_{esc} is equal to or greater than c, and thus the radius of a black hole must be less than or equal to

$$R = \frac{2GM}{c^2} \qquad (26.10)$$

This radius is called the **Schwartzschild radius**. The boundary of a sphere of radius R defines what is known as the **event horizon**. For example, if Earth were to become a black hole, its size would be reduced to that of a walnut (less than 10 mm).

*26.6 Relativistic Velocity Addition

The simple velocity addition formula, $u = v + u'$, holds only for objects moving at speeds much less than the speed of light.

Einstein derived the correct formula for adding velocities that is valid for all speeds from zero to the speed of light. If object 2 moves with a velocity u' relative to object 1, the velocity of object 2 relative to the observer, v, is given by

$$u = \frac{v + u'}{1 + \dfrac{vu'}{c^2}} \qquad (26.11)$$

where v is the velocity of object 1 with respect to the observer.

If v and u' are less than c, their sum v is then also less than c. If v or $u' = c$, $v = c$. If, $v = u' = c$, then $v = c$.

319

2007 Pearson Education, Inc., Upper Saddle River, NJ. All rights reserved. This material is protected under all copyright laws as they currently exist. No portion of this material may be reproduced, in any form or by any means, without permission in writing from the publisher.

Thus, the velocity addition formula is consistent with the constancy of the speed of light postulate.

Hints and Suggestions for Solving Problems

1. Use the relation $\Delta t = \dfrac{\Delta t_0}{\sqrt{1 - v^2/c^2}} = \gamma \, \Delta t_0$ for time dilation and the relation $L = L_0 \sqrt{1 - \dfrac{v^2}{c^2}} = L_0/\gamma$ for contracted length to solve problems in Section 26.3.

2. Remember that the factor $\sqrt{1 - \dfrac{v^2}{c^2}}$ is always less than

    1. Be careful, it is common to forget either to square the factor v/c or to take the square root. The factor tends to 1 as the velocity, v, tends to zero. As a result, the relativistic result will agree with the classical result when v is small.

3. Read the problem carefully several times to find out which observer sees the length contracted (or time dilated), and which observer sees its proper length (or proper time) before making an attempt to solve the problem. The hard part of solving problems on length contraction or time dilation is to figure out which length is proper length (or proper time) and which length is contracted length (or dilated time).

© 2007 Pearson Education, Inc., Upper Saddle River, NJ. All rights reserved. This material is protected under all copyright laws as they currently exist.
No portion of this material may be reproduced, in any form or by any means, without permission in writing from the publisher.

4. Motion in the perpendicular direction does not have any effect on the length.

5. When you add two velocities, remember that the result can never exceed c.

6. The relativistic momentum is γ times the classical momentum.

7. Remember that for velocities much less than c, relativistic results must agree with classical results.

2007 Pearson Education, Inc., Upper Saddle River, NJ. All rights reserved. This material is protected under all copyright laws as they currently exist. No portion of this material may be reproduced, in any form or by any means, without permission in writing from the publisher.

CHAPTER 27
QUANTUM PHYSICS

This chapter introduces the particle-like behavior of radiation. It describes blackbody radiation, the photoelectric effect, the Compton effect, and, the Bohr theory of the hydrogen atom, all of which can be satisfactorily explained only by the particle-like behavior of radiation. It also describes the laser and its applications.

Key Terms and Concepts

Thermal radiation
Blackbody
Wien's displacement law
Energy quantization
Ultraviolet catastrophe
Planck's hypothesis
Photon
Photoelectric effect
Stopping potential
Work function
Threshold frequency
Compton effect
Compton wavelength
Dual nature of light
Emission and absorption spectra
Balmer series
Principal quantum number
Laser
Metastable states
Stimulated emission
Popular inversion
Holography

322

© 2007 Pearson Education, Inc., Upper Saddle River, NJ. All rights reserved. This material is protected under all copyright laws as they currently exist. No portion of this material may be reproduced, in any form or by any means, without permission in writing from the publisher.

The classical theory of light could not satisfactory explain several phenomena of light discovered in the early twentieth century, which led to the idea that light is *quantized*. This hypothesis led to the formulation of new principles and a new branch, called *quantum mechanics*.

27.1 Quantization: Planck's Hypothesis

Hot objects emit electromagnetic radiation, called **thermal radiation**, that is proportional to the fourth power of their Kelvin temperatures.

At normal temperatures, the dominant radiation is in the infrared. At temperatures about 1000 K, the dominant glow is reddish and at temperatures of about 2000 K, it is yellowish white (such as the glowing filament of a light bulb). However, at any temperature the radiated energy forms a continuous spectrum. A **blackbody** is a substance that absorbs (and emits) all radiation incident on it.

A cavity with a very small opening, so that light can enter the cavity through the opening and be reflected many times inside until it is totally absorbed, behaves like a blackbody.

The distribution of energy in blackbody radiation has these two characteristics:
1. As the temperature increases, the total energy for each wavelength increases with temperature.
2. As the absolute temperature, T, is increased, the maximum intensity shifts toward increasing temperature. The wavelength (λ_m) at which the intensity is maximum is given by **Wien's displacement law**:

©2007 Pearson Education, Inc., Upper Saddle River, NJ. All rights reserved. This material is protected under all copyright laws as they currently exist. No portion of this material may be reproduced, in any form or by any means, without permission in writing from the publisher.

$$\lambda_m T = 2.90 \times 10^{-3}\,\text{m·K} \qquad (27.1)$$

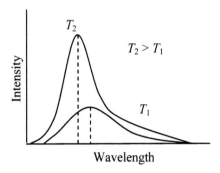

Wien's law is used to determine the temperatures of stars or other hot objects.

The Ultraviolet Catastrophe and Planck's Hypothesis
Classical physics was unsuccessful in explaining the distribution of blackbody radiation, particularly at short wavelengths. The theory predicted that at short wavelengths a blackbody would radiate infinite energy. This unrealistic result from classical theory at short wavelengths is known as the *ultraviolet catastrophe*.

To explain blackbody radiation, Planck hypothesized that the radiation energy of a thermal oscillator at a frequency f is an integral multiple of hf, where h is a universal constant, called **Planck's constant**, of value $h = 6.63 \times 10^{-34}$ J·s. Thus, the energy, E_n, of a thermal oscillator is

$$E_n = n(hf) \qquad n = 0, 1, 2, \dots \qquad (27.2)$$

324

© 2007 Pearson Education, Inc., Upper Saddle River, NJ. All rights reserved. This material is protected under all copyright laws as they currently exist. No portion of this material may be reproduced, in any form or by any means, without permission in writing from the publisher.

Equation (27.2) is called **Planck's hypothesis**, according to which the energy is **quantized**. The number n is called the quantum number, and the fundamental increment of energy, hf, is called a **quantum**. When $n = 1$

$$E_1 = hf \qquad (27.3)$$

For macroscopic systems, the quantum number n is very large, and a change of one quantum number is insignificant. The change in energy from one quantum state to the next is extremely small, and as a result their energies seem to change continuously. Energy jumps are important in atomic systems.

27.2 Quanta of Light: Photons and the Photoelectric Effect

Einstein introduced the concept of energy quantization of light, according to which the energy of emitted light (or any radiation) is quantized. A quantum or packet of light is called a **photon**. The energy of a photon is proportional to the frequency of the light. For a light of frequency f the energy of each photon is given by

$$E = hf \qquad (27.4)$$

where h is Planck's constant.

In the **photoelectric effect**, electrons are emitted from a substance when light shines on it. The ejected electrons are called *photoelectrons*. When a monochromatic light falls on a photocell, for positive anode voltages, photoelectric current does not vary with increasing voltage after all the emitted electrons reach the anode. For negative anode voltages, photoelectric current decreases with increasing negative voltage. At some voltage V_0, called the stopping potential, the current reduces to zero.

325

2007 Pearson Education, Inc., Upper Saddle River, NJ. All rights reserved. This material is protected under all copyright laws as they currently exist. No portion of this material may be reproduced, in any form or by any means, without permission in writing from the publisher.

The maximum kinetic energy (K_{max}) is related to the stopping potential by

$$K_{max} = eV_0 \qquad (27.5)$$

The maximum kinetic energy of the emitted electrons was found to depend linearly with the frequency of the incident light. No photoelectric effect was observed below a certain cutoff frequency f_0.

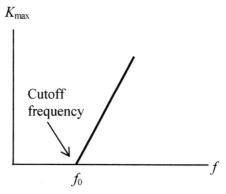

According to Einstein, in the photoelectric effect an incident light photon collides elastically with an electron in the material. Electrons are emitted if the photon has sufficient energy. The excess energy of the photon goes into kinetic energy of the ejected electron. From energy conservation,

$$hf = K + \phi \qquad (27.6)$$

where ϕ is the energy required to free an electron. The minimum energy needed to eject an electron from a material is called its **work function**, ϕ_0. Thus,

$$hf = K_{max} + \phi_0 \qquad (27.7)$$

326

© 2007 Pearson Education, Inc., Upper Saddle River, NJ. All rights reserved. This material is protected under all copyright laws as they currently exist. No portion of this material may be reproduced, in any form or by any means, without permission in writing from the publisher.

The cutoff frequency for the photoelectric effect is given by

$$f_0 = \frac{\phi_0}{h} \qquad (27.8)$$

f_0 is also called the **threshold frequency**.

If the intensity of the incident light increases without a change in its frequency, the maximum kinetic energy of the ejected electrons remains the same because the energy of each photon remains the same.

The photoelectric effect has practical applications such as in electric-eye door openers, automatic switches to turn on streetlights, laser scanners, and remote controls of TVs and VCRs. Photocells are used in solar energy to get electricity from sunlight.

27.3 Quantum "Particles": The Compton Effect

When a monochromatic X-ray beam is scattered by an electron, the scattered beam has a longer wavelength than the incident photon. This effect is called the **Compton effect**.

According to the wave theory, the scattered radiation should have the same wavelength. This is because the electrons in the scattering material absorb the incident radiation and oscillate at the same frequency to reemit the scattered radiation. According to the photon theory, the Compton effect is a consequence of energy conservation and momentum conservation in the collision between the

2007 Pearson Education, Inc., Upper Saddle River, NJ. All rights reserved. This material is protected under all copyright laws as they currently exist. No portion of this material may be reproduced, in any form or by any means, without permission in writing from the publisher.

incident X-ray photon (of energy $E = hf$) and an electron within the scattered material.

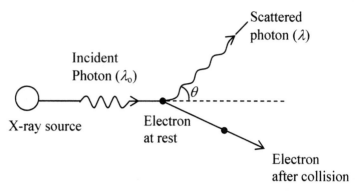

Applying these conservation principles, Compton showed that the change in wavelength, $\Delta\lambda$, in the scattering is

$$\Delta\lambda = \lambda - \lambda_o = \lambda_C (1 - \cos\theta) \qquad (27.9)$$

In this expression λ_o is the wavelength of the incident photon, λ is the wavelength of the scattered photon, m_e is the mass of electron, and θ is the angle between the incident photon and the scattered photon.

The quantity $\lambda_C = \dfrac{h}{m_e c} = 2.43 \times 10^{-12}$ m is called the

Compton wavelength of an electron.

The photoelectric effect and the Compton effect are strong evidence that support the photon nature of light.

© 2007 Pearson Education, Inc., Upper Saddle River, NJ. All rights reserved. This material is protected under all copyright laws as they currently exist. No portion of this material may be reproduced, in any form or by any means, without permission in writing from the publisher.

27.4 The Bohr Theory of the Hydrogen Atom

When heated, solid objects emit radiation that extends in a continuous distribution over all possible frequencies. This distribution is called a *continuous spectrum*. The radiation emitted by an isolated atom (such as from hydrogen gas in a low-pressure discharge tube) consists of a series of bright lines of different colors. The precise wavelength associated with each of these lines provides a way to identify atoms from their line spectra.

When atoms emit light, the observed spectrum is called an **emission spectrum**. Conversely, if light of all colors passes through a gas, the gas will absorb some wavelengths, giving rise to dark lines (where its atoms absorb colors) against the bright background, producing an **absorption spectrum** of the gas. The dark absorption lines appear at the same wavelengths as the emission lines.

The hydrogen atom is the simplest atom consisting of one electron and one proton. The following formula of Balmer gives the wavelength of the visible lines of the spectrum of hydrogen:

$$\frac{1}{\lambda} = R\left(\frac{1}{2^2} - \frac{1}{n^2}\right) \text{ for } n = 3, 4, 5, \dots \qquad (27.10)$$

The constant R is called the *Rydberg constant*. Its value is $1.097 \times 10^{-2} \text{ nm}^{-1}$. The series containing lines calculated using the preceding equation is called the **Balmer series**.

Bohr used the quantum concept of radiation to explain the line spectrum emitted by hydrogen atoms.

©2007 Pearson Education, Inc., Upper Saddle River, NJ. All rights reserved. This material is protected under all copyright laws as they currently exist. No portion of this material may be reproduced, in any form or by any means, without permission in writing from the publisher.

According to Bohr, the electron in an atom can rotate along certain selected circular orbits around the nucleus. For an orbiting electron, the electrostatic force between the electron and the nucleus (of charge Ze) provides the centripetal force.

$$\frac{mv^2}{r} = k\frac{e^2}{r^2} \qquad (27.11)$$

From this, he found the expression of total energy to be

$$E = -k\frac{e^2}{2r} \qquad (27.12)$$

When an electron is orbiting along any one of these orbits, its angular momentum is quantized. For the nth orbit, the angular momentum mvr is given by the relation

$$mvr = \frac{nh}{2\pi} \qquad (27.13)$$

where $n = 1, 2, 3, \ldots$, and h is Planck's constant. n is called the **principal quantum number**.

Using the preceding expression Bohr found that:

The speed of an electron when orbiting in the nth orbit is

$$v = \frac{nh}{2\pi mr} \qquad n = 1, 2, 3, \ldots$$

The radius of an electron orbiting in the nth orbit is

© 2007 Pearson Education, Inc., Upper Saddle River, NJ. All rights reserved. This material is protected under all copyright laws as they currently exist.
No portion of this material may be reproduced, in any form or by any means, without permission in writing from the publisher.

$$r_n = \left(\frac{h^2}{4\pi^2 k e^2 m}\right) n^2 \qquad (27.14)$$

The allowed energy of the nth orbit is given by

$$E_n = -\left(\frac{2\pi^2 k^2 e^4 m}{h^2}\right)\frac{1}{n^2} \qquad (27.15)$$

Using the numerical values of m, k, e, and h we get

$$r_n = (5.29 \times 10^{-11} \text{ m})n^2 \qquad (27.16)$$

$$E_n = -\frac{13.6}{n^2} eV \qquad (27.17)$$

Energy Levels
A plot of these energy values is known as an energy-level diagram. The lowest energy level ($n = 1$) is called the **ground state**. Higher energy levels are known as **excited states**. For example, the energy state corresponding to $n = 2$ is called the first excited state.

An electron is normally in the ground state and energy must be added to raise it to an excited state. An electron does not remain in the excited state for long. The time an electron spends in an excited state is called the **lifetime** of the state, which is typically about 10^{-8} s. According to Bohr, an atom gives off radiation when an electron jumps from a higher excited state to a lower state.
When an electron jumps from an energy state with $n = n_i$ to an energy state $n = n_f$,

331

©2007 Pearson Education, Inc., Upper Saddle River, NJ. All rights reserved. This material is protected under all copyright laws as they currently exist. No portion of this material may be reproduced, in any form or by any means, without permission in writing from the publisher.

$$\Delta E = 13.6 \left(\frac{1}{n_f^2} - \frac{1}{n_i^2} \right) \text{eV} \qquad (27.18)$$

This energy difference ΔE between the two states becomes the energy of the emitted light photon as the electron jumps from the higher state to the lower state. Thus, for an emitted photon, its frequency (f) (or the wavelength (λ) is given by $hf = hc/\lambda = \Delta E$.

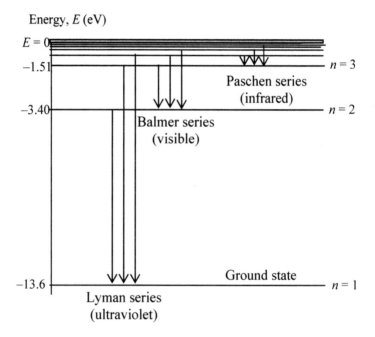

For the Balmer series (in the visible) $n_f = 2$. For the *Lyman series* (in the ultraviolet) $n_f = 1$. For the *Paschen series* (in the infrared) $n_f = 3$. In any such transition, the wavelength of the emitted photon is,

332

© 2007 Pearson Education, Inc., Upper Saddle River, NJ. All rights reserved. This material is protected under all copyright laws as they currently exis
No portion of this material may be reproduced, in any form or by any means, without permission in writing from the publisher.

$$\lambda = \frac{hc}{\Delta E} \qquad (27.19)$$

The wavelengths of light obtained from Bohr's theory agree well with the experimentally observed wavelengths of lines in the spectrum of hydrogen. Although the Bohr theory is successful in explaining the line spectrum of hydrogenic atoms it cannot successfully explain spectra of multielectron atoms. The theory is also incomplete since it introduces the quantum concept into the classical framework.

When the electrons in an atom are excited to higher-energy states, they may return to the ground state by a series of lower-energy jumps. These jumps produce lower-frequency (and, hence, longer-wavelength) photons than the photons that caused the original excitation. This process of emission of light of longer wavelength after illumination by shorter-wavelength light is called **fluorescence**. Absorbing ultraviolet light can excite atoms of some minerals. The minerals then glow by emitting a longer wavelength in the visible. Many small creatures such as corals and jellyfish show fluorescence.

27.5 A Quantum Success: The Laser

The word *laser* is an acronym for *l*ight *a*mplification by *s*timulated *e*mission of *r*adiation. Lasers produce light that is intense, highly directional, coherent (in phase), and pure in color.

According to the Bohr theory, when electrons drop from higher-energy states to lower-energy states of an atom, light photons are emitted. Emitted photons have random

333

2007 Pearson Education, Inc., Upper Saddle River, NJ. All rights reserved. This material is protected under all copyright laws as they currently exist. No portion of this material may be reproduced, in any form or by any means, without permission in writing from the publisher.

phase and they move in random directions. This is called spontaneous emission. In contrast, when an incident photon with energy equal to the energy difference between two states stimulates an electron in an atom to move to a lower-energy state, the resulting emission of a photon is called **stimulated emission**. In a stimulated emission, the emitted photon has the same energy and phase as the incident photon and both photons travel in the same direction. To produce stimulated emissions, the electrons must spend a longer time in the excited state than they would ordinarily so that incident photons continue to encounter excited atoms. A long-lived excited state is called a **metastable state**. **Phosphorescent** materials are examples of substances made of atoms with metastable states.

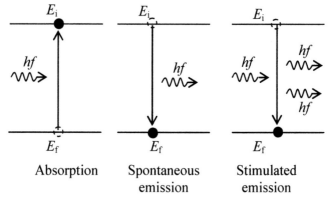

| Absorption | Spontaneous emission | Stimulated emission |

Repeated stimulated emissions increase the number of photons with the same energy and phase and traveling in the same direction, thus causing amplification of light. In order for light amplification to occur, there must be more atoms with electrons in the higher-energy orbit (excited state) than in the lower-energy orbit. This situation is known as **population inversion** in the system. Population

© 2007 Pearson Education, Inc., Upper Saddle River, NJ. All rights reserved. This material is protected under all copyright laws as they currently exist. No portion of this material may be reproduced, in any form or by any means, without permission in writing from the publisher.

inversion and stimulated emission are the two basic conditions necessary for laser action.

There are several types of lasers capable of producing light of different wavelengths. In a helium–neon laser, neon atoms are excited to a metastable state of energy 20.66 eV by resonant energy transfers from excited helium atoms. Because the lifetime of the excited state of helium is long (10^{-4} s), there is a good chance that it will collide with a neon atom before it spontaneously emits a photon. When the excited electrons of the neon atoms drop to energy level of energy 18.70 eV, they emit light photons of the same energy as the energy difference, 1.96 eV, between the levels, corresponding to the 632.8 nm red light of the laser.

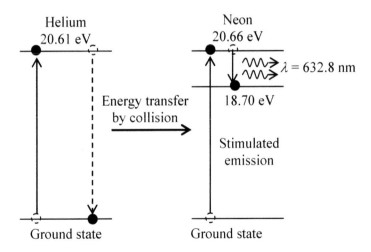

Helium
20.61 eV

Neon
20.66 eV

$\lambda = 632.8$ nm

Energy transfer
by collision

18.70 eV

Stimulated
emission

Ground state

Ground state

Lasers have many applications. In medicine, laser light is used to control bleeding, to weld torn retinas, and to treat cancers. The lasers that emit high-energy light photons are used in laser surgery to correct for nearsightedness and

335

2007 Pearson Education, Inc., Upper Saddle River, NJ. All rights reserved. This material is protected under all copyright laws as they currently exist.
No portion of this material may be reproduced, in any form or by any means, without permission in writing from the publisher.

farsightedness. In industry, lasers are used to drill and to weld. They are also used in surveying and to carry telephone and television signals over fiber optic cables. Lasers are used in laser printers, compact disc players, and in supermarket checkouts.

Lasers are used in three-dimensional imaging called **holography**. In this case, a laser beam is split into reference and object beams. The object beam falls on the object. The interference patterns of these two beams are recorded on a photographic plate. This developed plate is called a *hologram*. When the laser light illuminates the hologram, it shows a three-dimensional image of the object.

Hints and Suggestions for Solving Problems

1. Some useful constants for this chapter are h (Planck's constant) $= 6.63 \times 10^{-34}$ J·s; c (speed of light) $= 3 \times 10^{8}$ m/s, m (mass of an electron) $= 9.1 \times 10^{-31}$ kg, e (magnitude of the charge of an electron) $= 1.6 \times 10^{-19}$ C, and R (Rydberg constant) $= 1.097 \times 10^{-2}$ nm^{-1} $= 1.097 \times 10^{7}$ m^{-1}.

2. According to Wien's displacement law, $\lambda_m T = 2.90 \times 10^{-3}$ m·K. Be sure to convert the temperature to kelvins when using this law to calculate the temperature of a star or any blackbody, such as in the problems in Section 27.1.

3. Wavelength is usually given in nanometers (nm). Convert it to meters (m) when solving problems.

© 2007 Pearson Education, Inc., Upper Saddle River, NJ. All rights reserved. This material is protected under all copyright laws as they currently exist. No portion of this material may be reproduced, in any form or by any means, without permission in writing from the publisher.

4. The energy of a light photon is $E = hf = hc/\lambda$. Use this expression to calculate the frequency or wavelength of a photon.

5. Energy may be given in electron volts (eV) or in joules (J) in numerical problems. If the value of h to be used is in J·s, make sure to convert energy to joules. Remember the conversion $1 \text{ eV} = 1.6 \times 10^{-19}$ J.

6. The stopping potential in the photoelectric effect is related to the maximum energy of the photoelectrons by $K_{max} = eV_0$.

7. The cutoff frequency or cutoff wavelength for the photoelectric effect can be obtained from the expression $\phi_0 = E = hf_0 = hc/\lambda_0$, where ϕ_0 is called the work function. Usually it is given in eV. Be sure to convert it to joules (J). The maximum kinetic energy of a photoelectron is $KE_{max} = hf - \phi_0$.

8. The cutoff frequency is $f_0 = \phi_0/h$. Use tips 6, 7, and 8 to solve problems in Section 27.2.

9. Calculation of the change of wavelength of an X-ray by Compton scattering is straightforward. Use the expression $\Delta\lambda = \lambda' - \lambda = (h/m_e c)(1 - \cos\theta)$ for the change in wavelength or the scattered wavelength. The constant $h/m_e c$ is 2.43×10^{-12} m. Also, the momentum of a photon is $p = hf/c = h/\lambda$. The preceding expressions will be useful for solving problems in Section 27.3.

10. For a hydrogen atom, the radius of its nth orbit is $r_n = 0.0529\, n^2$ nm $= (5.29 \times 10^{-11}$ m$)n^2$, and the energy of its nth orbit is $E_n = -13.6\,/n^2$ eV. Use the preceding

337

2007 Pearson Education, Inc., Upper Saddle River, NJ. All rights reserved. This material is protected under all copyright laws as they currently exist. No portion of this material may be reproduced, in any form or by any means, without permission in writing from the publisher.

expression for energy to determine the energy difference ΔE between the initial state and final state when an electron jumps from one energy state to the other. Set the energy difference to hc/λ to calculate the wavelength of light emitted. Remember to convert electron volts to joules. The preceding expressions and technique will be useful for solving problems in Section 27.4.

11. Remember that for the Balmer series in the hydrogen spectrum (in the visible) $n_f = 2$. For the *Lyman series* (in the ultraviolet) $n_f = 1$. For the *Paschen series* (in the infrared) $n_f = 3$.

© 2007 Pearson Education, Inc., Upper Saddle River, NJ. All rights reserved. This material is protected under all copyright laws as they currently exist. No portion of this material may be reproduced, in any form or by any means, without permission in writing from the publisher.

CHAPTER 28
QUANTUM MECHANICS AND ATOMIC PHYSICS

This chapter introduces the particle-like behavior of radiation and basic ideas of quantum mechanics. It describes matter waves, the Schrödinger wave equation, atomic quantum numbers, and the Heisenberg uncertainty principle, as well as particles and antiparticles.

Key Terms and Concepts

Matter wave
De Broglie hypothesis
Wave function
Schrödinger wave equation
Probability density
Tunneling
Orbital quantum number
Magnetic quantum number
Spin quantum number
Pauli exclusion principle
Period table
The Heisenberg uncertainty principle
Positron
Antiparticle
Pair annihilation
Antimatter

28.1 Matter Waves: The De Broglie Hypothesis

According to relativity, a light photon is a massless particle that travels with the speed of light and its energy is

2007 Pearson Education, Inc., Upper Saddle River, NJ. All rights reserved. This material is protected under all copyright laws as they currently exist. No portion of this material may be reproduced, in any form or by any means, without permission in writing from the publisher.

$E = p\,c$, p being the momentum of the photon. Also, for a photon $E = hf = hc/\lambda$. Thus, for a photon,

$$p = \frac{E}{c} = \frac{h}{\lambda} \tag{28.1}$$

According to Louis de Broglie, since light, which is usually treated as a wave, shows particle-like behavior, particles of matter (such as the electron) should exhibit wavelike behavior.

According to the de Broglie hypothesis, *a particle of momentum of magnitude p has a wave associated with it*, the wavelength of which is given by

$$\lambda = \frac{h}{p} = \frac{h}{mv} \tag{28.2}$$

The waves associated with moving particles are called **matter waves** or **de Broglie waves**. The de Broglie wavelength is extremely small for macroscopic objects and may be unobservable. For a microscopic system, such as the motion of an electron, the de Broglie wavelength is of the order of the size of atoms and molecules and thus is important in such systems.

In support of his hypothesis, de Broglie showed that the allowed orbits of electrons in a hydrogen atom correspond to the formation of standing wave patterns by the electrons. The circumference of the Bohr orbit is equal to the integer multiple of the wavelength of the electron in that orbit. From these considerations, de Broglie derived the angular momentum condition, $L_n = mvr_n = nh/2\pi$ ($n = 1, 2, 3\ldots$) for the motion of an electron in that orbit. Thus,

340

© 2007 Pearson Education, Inc., Upper Saddle River, NJ. All rights reserved. This material is protected under all copyright laws as they currently exist. No portion of this material may be reproduced, in any form or by any means, without permission in writing from the publisher.

he showed that angular momentum is quantized, as Bohr proposed.

The success of de Broglie's matter waves in deriving Bohr's angular momentum condition led to the development of Schrödinger's wave equation and the probabilistic interpretation of the amplitude of a matter wave.

Electrons can have wavelengths of the order of the size of atoms in a crystal. As a result they show diffraction effects when reflected from the atomic layers of a crystal. In 1927, Davison and Germer had shown the diffraction of electrons by a nickel crystal. The agreement between the experimental value of the wavelength of the electron with the value predicted from the de Broglie hypothesis provided a convincing proof of the matter wave.

If an electron is accelerated through a potential difference, V, its kinetic energy, $p^2/2m = eV$. This gives the de Broglie wavelength

$$\lambda = \frac{h}{p} = \sqrt{\frac{150}{V}} \text{ nm.} \qquad (28.3)$$

A practical application of de Broglie waves is seen in the electron microscope. In an electron microscope, a beam of electrons is used instead of light for imaging. The wavelength associated with the beam of electrons is typically 1000 times shorter than the wavelength of visible light. Since resolution depends inversely on wavelength, an electron microscope has more resolution and thus can show much finer detail than an optical microscope. There are two kinds of electron microscopes: *transmission electron microscope* and *scanning electron microscope*.

341

2007 Pearson Education, Inc., Upper Saddle River, NJ. All rights reserved. This material is protected under all copyright laws as they currently exist. No portion of this material may be reproduced, in any form or by any means, without permission in writing from the publisher.

A scanning tunneling microscope (STM) works using quantum tunneling. The tunneling current, which depends on the gap between the specimen and the tip, is kept constant. As the surface is scanned the tip moves up and down to keep the current constant. The movement of the tip is converted into the visual image of the surface showing the peaks and valleys of the atoms on the surface.

28.2 Schrödinger Wave Equation

According to de Broglie, moving particles have associated waves that describe their behavior. The de Broglie wave is described by a mathematical function, called the wave function, ψ, that describes the wave as a function of space and time. The wave function is associated with the particle's energy (both kinetic and potential) and it follows an equation:

$$(K + U)\, \psi = E\, \psi \qquad (28.4)$$

The preceding equation is known as the **Schrödinger wave equation**.

The solution of Schrödinger's equation gives a matter wave, or wave function, corresponding to a particular physical system. In general, ψ varies in magnitude with space and time. For a single particle, ψ^2 represents the probability of finding the particle at various locations. If the wave function ψ represents a beam of particles, then ψ^2 will be proportional to the number of particles in a small volume at a given location at a particular time.

In this probabilistic description of ψ, a probability cloud represents an electron in an atom, with a denser cloud

342

© 2007 Pearson Education, Inc., Upper Saddle River, NJ. All rights reserved. This material is protected under all copyright laws as they currently exist. No portion of this material may be reproduced, in any form or by any means, without permission in writing from the publisher.

representing a region of higher probability of finding the electron. Thus, an electron in a hydrogen atom does not always move in a circle at a particular distance from the nucleus. The electron can be found at any distance, with its maximum probability that it will be found at a distance equal to the Bohr radius (0.0529 nm) of the hydrogen atom.

28.3 Atomic Quantum Numbers and the Periodic Table

The Hydrogen Atom

The complete picture of a hydrogen atom comes from Schrödinger's wave equation, which predicted the same allowed energy levels as found from the Bohr theory. The energy value depended only on the principal quantum number n. However, the solution gave two other quantum numbers, the orbital quantum number, l, and the magnetic quantum number, m_l. In fact, in addition to the principal quantum number there are three more quantum numbers that are associated with an electron in a hydrogen atom, as described below.

The **orbital quantum number** is associated with the orbital angular momentum of an electron. The orbital quantum number l can have the values $l = 0, 1, 2, \ldots$ $(n - 1)$. The orbits (such as those in a hydrogen atom) with the same value of n but different values of l are called *degenerate*.

The **magnetic quantum number** is associated with the orientation of the orbital angular momentum vector in space. In the absence of a magnetic field all orientations of the angular momentum vector have the same energy. m_l is introduced to determine the number of levels that exist

343

©2007 Pearson Education, Inc., Upper Saddle River, NJ. All rights reserved. This material is protected under all copyright laws as they currently exist. No portion of this material may be reproduced, in any form or by any means, without permission in writing from the publisher.

for a given l. The allowed values of m_l are $m_l = -l, -l+1, -l+2, \ldots, -1, 0, +1, +2, \ldots, l-1, l$.

A high-resolution spectrometer found that each emission line of hydrogen is two closely spaced lines. This high resolution splitting is called *fine structure*. Thus, a fourth quantum number is needed for a given orbit to explain fine structure of spectral lines. The electron **spin quantum number**, m_s, is associated with the *intrinsic* angular momentum of an electron. It can have the values $m_s = -\frac{1}{2}$, $\frac{1}{2}$. These two values correspond to the electron's spin being "up" or "down" with respect to the z-axis, respectively. An electron's spin is fundamentally a pure quantum-mechanical property of the particle.

Multielectron Atoms

In multielectron atoms the electrons experience mutual repulsive electrostatic forces in addition to their attractive interactions with the nucleus. The Schrödinger equation for multielectron atoms can be solved only to a workable approximation. Application of the Schrödinger equation to multielectron atoms shows that the energy level of a multielectron atom depends on the principal quantum number, n, and the orbital quantum number, l. The energy value increases with n for a given l and also with l for a given n.

Electrons having the same value of n are said to be in the same **shell**. For $n = 1$, an electron is in the K shell, for $n = 2$ it is in the L shell, and for $n = 3$ it is in the M shell, and so on. Electrons in a given shell with the same value of l are said to be in the same **subshell**. The subshells are named $s, p, d, f, g \ldots$ for $l = 0, 1, 2, 3, 4 \ldots$, respectively. Since an electron's energy depends on both n and l, both quantum numbers are used to denote energy levels. For

344

© 2007 Pearson Education, Inc., Upper Saddle River, NJ. All rights reserved. This material is protected under all copyright laws as they currently exis[t]
No portion of this material may be reproduced, in any form or by any means, without permission in writing from the publisher.

example $1s$ denotes an energy level with $n = 1$ and $l = 0$. It is also customary to refer to the m_l values as representing orbitals. For example, a $2p$ energy level has three orbitals for $m_l = 1, 0, -1$. The shell-subshell (n-l) sequence shows that for multielectron atoms the energy levels are not evenly spaced and the sequence of energy levels has numbers out of order. For example, the $4s$ energy level lies below the $3d$ level. Such variations result in part from electrical forces between the electrons in the atom. The maximum number of electrons in a subshell is $2(2l+1)$.

The Pauli Exclusion Principle

In a multielectron atom it is not possible for all the electrons to exist in the lowest-energy state. Only one electron at a time may have a particular set of quantum numbers n, l, m_l, and m_s. The **Pauli exclusion principle** states: *No two electrons in a multielectron atom can have the same set of quantum numbers (n, l, m_l, and m_s).* That is, once an electron occupies a particular state, other electrons are excluded from that state.

The limit on the quantum numbers sets the limits on the number of electrons that can occupy an energy state. For example the $1s$ state ($n = 1$, $l = 0$) can have only two electrons with two unique sets of quantum numbers (1, 0, 0, +1/2) and (1, 0, 0, -1/2). A shorthand notation called the **electronic configuration** usually describes the electron occupancy of different energy levels. For example, the element lithium, which has two electrons in the $n = 1$, $l = 0$ state and one electron in the $n = 2$ state, has the electronic configuration $1s^2 2s^1$. Neon, with ten electrons, has the electronic configuration $1s^2 2s^2 2p^6$. The energy spacing between adjacent subshells is not uniform. The sets of energy levels with about the same energy are called **electron periods**.

345

2007 Pearson Education, Inc., Upper Saddle River, NJ. All rights reserved. This material is protected under all copyright laws as they currently exist. No portion of this material may be reproduced, in any form or by any means, without permission in writing from the publisher.

The Periodic Table of Elements
The Russian chemist Mendeleev arranged all the chemical elements in a table, called the **periodic table of elements**, by grouping the elements based on their chemical properties. Mendeleev arranged elements in rows, called **periods** and vertical columns called **groups**. The periodic table puts the elements into seven periods in order of increasing atomic number. As electrons fill subshells of increasing l they produce different elements of the periodic table. The first period has only two elements, periods 2 and 3 have eight elements, and periods 4 and 5 have eighteen elements. In *representative elements* the last electron enters an s or p subshell. In *transition elements* the last electron enters a d subshell whereas in inner transition elements the last electron enters an f subshell.

Atoms with the same configuration of the outermost electrons have similar chemical properties. For example, lithium, sodium, potassium, rubidium, cesium, and francium belong to the same group and are called alkali metals. They have similar chemical properties, and all have only one electron in the outermost subshell. The electrons, called *valence electrons*, in the outermost *unfilled* subshell determine the chemical properties of an element.

28.4 The Heisenberg Uncertainty Principle

In classical mechanics there is no limit to the accuracy of a measurement. A natural consequence of the wave properties of a particle is that the position and momentum of a particle cannot be simultaneously determined with arbitrary accuracy. That is, if the position of a particle is known with a greater accuracy, its momentum becomes

© 2007 Pearson Education, Inc., Upper Saddle River, NJ. All rights reserved. This material is protected under all copyright laws as they currently exist. No portion of this material may be reproduced, in any form or by any means, without permission in writing from the publisher.

uncertain, and vice versa. This consequence is expressed in the **Heisenberg uncertainty principle** as:

It is impossible to know simultaneously an object's exact position and momentum. Heisenberg showed that:

$$\Delta p_x \, \Delta x \geq \frac{h}{2\pi} \tag{28.5}$$

where Δx is the uncertainty in the measurement of the position, x, and Δp_x is the uncertainty in the measurement of the corresponding momentum.

The inaccuracy of simultaneous measurements of energy, E, and time, t, is expressed in the Heisenberg uncertainty principle of energy and time as:

$$\Delta E \, \Delta t \geq \frac{h}{2\pi} \tag{28.6}$$

Because a measurement of energy needs to be made in a time comparable to the lifetime (Δt) of an electron in an excited state, the energy of that state is uncertain by an amount $\Delta E = h\Delta f$. Because of this the observed line in the emission spectrum of an atom has a finite width ΔE. This is called the *natural broadening* of a line.

28.5 Particles and Antiparticles

In 1928 Dirac extended the quantum theory of Schrödinger and Heisenberg to include relativistic considerations. He predicted from his relativistic quantum theory the existence of a particle having the same mass as that of an electron but carrying a *positive* charge. This particle is

2007 Pearson Education, Inc., Upper Saddle River, NJ. All rights reserved. This material is protected under all copyright laws as they currently exist. No portion of this material may be reproduced, in any form or by any means, without permission in writing from the publisher.

called the **positron** and is considered to be the **antiparticle** of an electron.

Anderson experimentally discovered the positron in 1932 in a cloud-chamber experiment with cosmic rays from outer space containing highly energetic X-rays. In the experiment, an X-ray photon interacted with an atomic nucleus of lead to produce an electron and a positron. This process is called **pair production**. An external field deflected the electron and the positron in paths of opposite curvatures.

Because of the conservation of electric charge, a positron is always created with the simultaneous creation of an electron. The *minimum* photon energy, called the **threshold energy** for pair production, to produce an electron–positron pair is:

$$E_{\min} = hf = 2m_e c^2 = 1.022 \text{ MeV} \qquad (28.7)$$

An energetic positron produced in a pair production loses its energy as it passes through matter and finally combines with an electron of that material to form a *positronium atom*, which is a hydrogen-like atom with a positron substituted for a proton. The positronium atom is unstable and decays quickly (in about 10^{-10} s) into two photons each with an energy of 0.511 MeV. This process, which is a direct conversion of mass into electromagnetic energy, is called **pair annihilation**.

All subatomic particles have antiparticles that can be created by cosmic rays entering our atmosphere and/or by nuclear processes. It is possible that **antiparticles** predominate in some parts of the universe. If so, the atoms of the antimatter would consist of negatively charged

© 2007 Pearson Education, Inc., Upper Saddle River, NJ. All rights reserved. This material is protected under all copyright laws as they currently exist. No portion of this material may be reproduced, in any form or by any means, without permission in writing from the publisher.

nuclei containing antiprotons and antineutrons surrounded by orbiting positrons. The physical behavior of antimatter atoms would likely be the same as ordinary matter atoms. However, when antimatter and matter come in contact, they would annihilate each other liberating a tremendous amount of energy.

Hints and Suggestions for Solving Problems

1. The de Broglie wavelength associated with a particle is $\lambda = h/p$, where $p = mv$. Use this relation to determine wavelength (λ), momentum (p), or velocity (v) of the particle. Once v is known, the kinetic energy of the particle can be calculated from its mass, m, and velocity, v.

2. Remember that if an electron is accelerated by a potential difference V, its energy is $1/2mv^2 = p^2/2m = Ve$. This relation together with $\lambda = h/p$, gives $\lambda = \sqrt{150/V}$ nm. Tips 1 and 2 will be useful to solve problems in Section 28.1.

3. Remember that for a given value of the principal quantum number, n, the values of the orbital angular momentum quantum number, l, are 0, 1, 2, ..., $(n-1)$. The allowed values of the magnetic quantum numbers, m_l, are $-l, -l+1, -l+2, ..., -1, 0, +1, +2, ..., l-1, l$. The spin quantum number, m_s, can have two values $-\frac{1}{2}, \frac{1}{2}$.

4. Use the Heisenberg uncertainty relation $\Delta p_x \, \Delta x \geq h/2\pi$ or $\Delta E \, \Delta t \geq h/2\pi$ to determine the

349

2007 Pearson Education, Inc., Upper Saddle River, NJ. All rights reserved. This material is protected under all copyright laws as they currently exist. No portion of this material may be reproduced, in any form or by any means, without permission in writing from the publisher.

uncertainties relating to the simultaneous measurements of position and momentum or energy and time, respectively, such as in the problems in Section 28.4.

© 2007 Pearson Education, Inc., Upper Saddle River, NJ. All rights reserved. This material is protected under all copyright laws as they currently exist. No portion of this material may be reproduced, in any form or by any means, without permission in writing from the publisher.

CHAPTER 29
THE NUCLEUS

This chapter focuses on the nucleus of an atom—its composition and stability. The forces inside a nucleus and radioactive decays are described. Practical applications, detection, and biological effects of radiation are also discussed.

Key Terms and Concepts

Plum-pudding model
Rutherford-Bohr model
Atomic number
Strong nuclear force
Mass number and neutron number
Isotopes
Nuclide
Radioactivity
Alpha and beta particles
Gamma rays
Alpha decay, beta decay, and gamma decay
Tunneling or barrier penetration
Electron capture
Activity
Half-life
Radioactive dating
Carbon-14 dating
Binding energy
Magic numbers
Radiation detector
Roentgen (R)
Gray (G)
Rem
Relative biological effectiveness (RBE)

351

2007 Pearson Education, Inc., Upper Saddle River, NJ. All rights reserved. This material is protected under all copyright laws as they currently exist. No portion of this material may be reproduced, in any form or by any means, without permission in writing from the publisher.

29.1 Nuclear Structure and the Nuclear Force

It is evident from the photoelectric effect and *thermionic emission* (emission of electrons from heated elements) that an atom contains negatively charged electrons. An atom, being electrically neutral, therefore includes both the electrons and an equal number of positive charges. According to the "plum-pudding" model of Thomson, the positive charges are uniformly distributed within an atom, whereas the electrons are symmetrically situated inside the atom.

The modern model of atomic structure is based on Rutherford's alpha particle scattering experiment by atoms. Combined with the Bohr theory, this model is called the **Rutherford-Bohr model** of the atom. An alpha particle is a doubly positively charged particle emitted by some radioactive nuclei. By directing a beam of positively charged alpha particles on a thin foil of gold in his experiment, Rutherford found that

(i)   Most alpha particles went straight without being deflected, and

(ii)   Only some alpha particles were deflected at very large angles.

According to Thomson's plum pudding model of an atom, the massive alpha particles are expected to be deflected only slightly by collision with the electrons in the atom. To account for the results of his experiment, Rutherford declared that all positive charges in an atom are concentrated at the center, called the **nucleus**, of an atom,

352

© 2007 Pearson Education, Inc., Upper Saddle River, NJ. All rights reserved. This material is protected under all copyright laws as they currently exist.
No portion of this material may be reproduced, in any form or by any means, without permission in writing from the publisher.

whereas for the most part an atom is empty. The negatively charged electrons rotate around the nucleus in circular orbits. An atom thus has a structure similar to the solar system.

For a head-on collision between an incoming alpha particle of energy $\frac{1}{2}mv^2$ and the nucleus, if r_{min} is the distance of closest approach of the alpha particle to the nucleus, the alpha particle will be stopped at this distance by the repulsive Coulomb force (between the alpha particle and the nucleus) and will be accelerated backward. By work–energy theorem, the work done W by the Coulomb force ($W = (2e)V$ where $V = kZe/r$) is equal to the kinetic energy of the incoming alpha particle. From this consideration, Rutherford found that:

$$r_{min} = \frac{4kZ\,e^2}{m\,v^2} \qquad (29.1)$$

Here Z is the atomic number, or the number of protons in the nucleus, and Ze is the total nuclear charge. r_{min} is found to be about 10^{-14} m, which is the upper limit for the radius of a nucleus.

In 1932, another particle, called the neutron, was discovered from the nucleus of an atom. The nucleus of an atom is thus composed of two types of particles: protons and neutrons. A neutron is electrically neutral and has a mass slightly greater than that of a proton. These two types of particles are collectively called **nucleons**.

The Nuclear Force
The protons inside a nucleus repel one another because of the Coulomb force. The gravitational attractive force between nucleons is negligible compared to the repulsive

353

2007 Pearson Education, Inc., Upper Saddle River, NJ. All rights reserved. This material is protected under all copyright laws as they currently exist. No portion of this material may be reproduced, in any form or by any means, without permission in writing from the publisher.

Coulomb force. Since most atoms are stable, there must be a large attractive force, called the **strong nuclear force**, among the nucleons in a nucleus that holds the nucleus together. The strong force is short range, and extends over a distance of only 10^{-15} m. It is much stronger than the Coulomb force and acts between any two nucleons (that is, between two protons, two neutrons, and between a proton and neutron).

Nuclear Notation
The atomic number, Z, is the number of protons in a nucleus. This number is called the **proton number**. The **mass number**, A, of a nucleus is the number of nucleons in the nucleus. Thus, the neutron number, N, of a nucleus is $N = A - Z$. A complete representation of the nucleus is expressed by the chemical symbol of the element with the mass number A as a left superscript and the proton number Z as a left subscript. Thus, a nucleus X with proton number Z and mass number A is written as $^A_Z X$. For example, the nucleus of carbon-14 (which has 6 protons and 8 neutrons) is $^{14}_6 C$.

Nuclei with the same atomic number but different numbers of neutrons are called **isotopes**. A particular isotope of any element is called a **nuclide**. Hydrogen has three isotopes or nuclides. The isotope ^1H is called hydrogen, ^2H is called *deuterium* (or heavy water), and the isotope ^3H is called *tritium* (which is unstable).

29.2 Radioactivity

Radioactivity refers to the emissions from an unstable nucleus. In radioactivity, the nucleus changes its

354

© 2007 Pearson Education, Inc., Upper Saddle River, NJ. All rights reserved. This material is protected under all copyright laws as they currently exist. No portion of this material may be reproduced, in any form or by any means, without permission in writing from the publisher.

composition by emitting particles, or an excited nucleus decays to a lower-energy state by emitting photons. Experiments showed that the radiation emitted by a radioactive nucleus is of three kinds containing the following three particles as distinguished by passing them through a magnetic field:

Alpha particles are doubly charged ($+2e$) particles containing two protons and two neutrons. That is, these are nuclei of $_2^4 \text{He}$.

Beta particles are electrons.

Gamma particles are photons.

Alpha Decay

A nucleus that emits an alpha particle loses two neutrons and two protons. That is, it decreases its mass by 4 and its proton number by 2. This decay is represented as

$$_Z^A \text{X} \longrightarrow \quad _{Z-2}^{A-4} \text{Y} + _2^4 \text{He}$$

where X and Y are the parent and daughter nuclei, respectively.

An example of alpha decay is:

$$_{84}^{214} \text{Po} \longrightarrow _{82}^{210} \text{Pb} + _2^4 \text{He}$$

Conservation of nucleons (according to which the total number of nucleons remains constant) and **conservation of charge** (according to which the total charge remains constant) apply in any nuclear process.

The energies of the alpha particles emitted from radioactive sources are typically a few MeV. For example, in the decay of a ^{238}U nucleus the emitted alpha particles have energy of 4.14 MeV. Classically, these alpha particles do not have enough energy to overcome the

355

©2007 Pearson Education, Inc., Upper Saddle River, NJ. All rights reserved. This material is protected under all copyright laws as they currently exist. No portion of this material may be reproduced, in any form or by any means, without permission in writing from the publisher.

electrostatic potential barrier of the ^{238}U nucleus and are scattered.

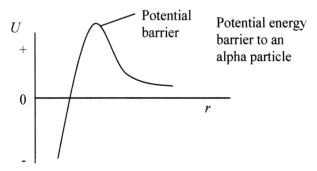

Quantum mechanics predicts a finite probability of finding a particle in a classically forbidden region if the wave function exists in the region. The square of the alpha particle wave function gives the probability of finding it in a given location. Quantum mechanically, inside the nucleus the probability amplitude for the alpha particle is large (that is, there is a high likelihood of its being found inside the nucleus). But it can "tunnel" through the potential barrier and appear outside the nucleus with a smaller, but nonzero and finite, probability. This is called quantum **tunneling** or **barrier penetration**.

Beta Decay
In **beta decay**, an electron is emitted from the nucleus. It indicates *that an electron is created in the nucleus* before it is emitted. This process is represented by

$$_Z^A X \; \longrightarrow \; _{Z+1}^A Y + _{-1}^0 e$$

For example, a basic β^- **decay** is:

$$_6^{14} C \; \longrightarrow \; _7^{14} N + _{-1}^0 e$$

356

© 2007 Pearson Education, Inc., Upper Saddle River, NJ. All rights reserved. This material is protected under all copyright laws as they currently exist. No portion of this material may be reproduced, in any form or by any means, without permission in writing from the publisher.

In this decay, a neutron is converted to a proton and an electron:

$$\ _0^1 n \ \longrightarrow \ \ _1^1 p \ + \ _{-1}^0 e$$

Thus, the neutron number decreases by one and the proton number increases by one in a beta decay. It is to be noted that another particle called an antineutrino is also emitted in beta decay.

Some unstable isotopes undergo *positron decay* or β^+ decay. This process involves the emission of a positron from the nucleus. A proton inside the nucleus is converted to a neutron in this process. This process is accompanied by a particle, called a *neutrino*. An example of β^+ **decay** is,

$$\ _8^{15} O \ \longrightarrow \ \ _7^{15} N \ + \ _{+1}^0 e$$

A third related process is called **electron capture (EC)**, which involves the absorption of *orbital* electrons by a nucleus. This process produces the same daughter nucleus as would have been produced in positron decay. An example of electron capture is:

$$\ _4^7 Be \ + \ _{-1}^0 e \ \longrightarrow \ \ _3^7 Li$$

Usually an electron from the innermost shell (called K shell) is captured. This is called *K-capture*. Electron capture in more distant orbits (such as *L-capture*) occurs with much less probability. An electron capture is detected by observing the characteristic X-ray emission

2007 Pearson Education, Inc., Upper Saddle River, NJ. All rights reserved. This material is protected under all copyright laws as they currently exist. No portion of this material may be reproduced, in any form or by any means, without permission in writing from the publisher.

when an orbiting electron from an outer shell makes a transition to the K-shell vacancy.

Gamma Decay

In a **gamma decay**, an excited nucleus decays to a lower-energy state by emitting high-energy photons. In this process, neither the mass number nor the proton number of the nucleus is changed. An example of gamma decay is:

$$^{61}_{28}\text{Ni}* \longrightarrow \ ^{61}_{28}\text{Ni} + \gamma \ (\text{gamma ray})$$

The asterisk indicates that the ^{61}Ni nucleus is in an excited state.

Radiation Penetration

Absorption or degree of penetration of nuclear radiation has important applications in the treatment of cancers or radioactive shielding.

Alpha particles are doubly charged, have a larger mass, and move slower. They can be stopped by a few cm of air or by a sheet of paper.

Beta particles are singly charged and less massive. They can travel a few meters in air or a few mm in aluminum before being stopped.

Gamma rays are neutral photons and more penetrating. A beam of high-energy gamma rays can penetrate a cm or more of lead.

Radiation can do damage to materials and living things. Materials may lose strength when exposed to such radiation. In biological tissue, radiation damage occurs

358

© 2007 Pearson Education, Inc., Upper Saddle River, NJ. All rights reserved. This material is protected under all copyright laws as they currently exist. No portion of this material may be reproduced, in any form or by any means, without permission in writing from the publisher.

due to ionization in living cells. The intensity of radiation from cosmic rays is usually too low to be harmful.

If the daughter nucleus of an unstable nucleus is also unstable, it will decay to produce its own daughter nucleus. As a result, an unstable parent nucleus will produce a series of radioactive nuclei, called a radioactive decay series. Of nearly 1200 known unstable nuclides only a small number occur naturally. Most of the radioactive nuclides found in nature occur as a product of the decay series of heavier nuclei.

29.3 Decay Rate and Half-Life

Radioactive nuclei decay with time in a well-defined manner. The number of decays per second is called the **activity** of a sample. The rate at which the number of parent nuclei (N) decreases is given by:

$$\frac{\Delta N}{\Delta t} = -\lambda N \tag{29.2}$$

where the constant λ is called the decay constant. The unit of λ is s^{-1}.

If N_0 is the number of nuclei at $t = 0$, the number of nuclei, N, remaining at time t is given from the preceding equation as:

$$N = N_0 e^{-\lambda t} \tag{29.3}$$

The plot of N/N_0 with t is shown.

2007 Pearson Education, Inc., Upper Saddle River, NJ. All rights reserved. This material is protected under all copyright laws as they currently exist. No portion of this material may be reproduced, in any form or by any means, without permission in writing from the publisher.

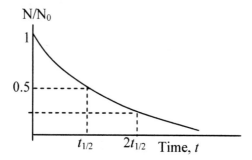

The **half-life** of a radioactive material is the time required for a given number of its nuclei to decrease to half their original number. In terms of the decay constant, the half-life is

$$t_{1/2} = \frac{\ln 2}{\lambda} = \frac{0.693}{\lambda} \qquad (29.4)$$

The strength of a radioactive sample at a given time is given by its activity. The units of activity are the **curie (Ci)** and the **becquerel (Bq)**.

$$1 \text{ Ci} = 3.70 \times 10^{10} \text{ decays/s}$$
$$1 \text{ Bq} = 1 \text{ decay/s}$$

Radioactive Dating

Because the decay rates of radioactive nuclides are constant, they can be used as nuclear clocks. Scientists use half-life to determine the age of objects containing radioactive nuclides. Carbon-14 is unstable. The carbon-14 activity of an organism is constant until it dies, at which point the activity decreases exponentially with a half-life of 5730 years. Carbon-14 can be used to date organic materials with ages of up to 10,000 to 15,000 years. The

360

© 2007 Pearson Education, Inc., Upper Saddle River, NJ. All rights reserved. This material is protected under all copyright laws as they currently exist. No portion of this material may be reproduced, in any form or by any means, without permission in writing from the publisher.

initial activity, R_0, of carbon-14, for a 1-g of sample of carbon is 0.230 Bq.

To measure dates on different time scales, different radioactive isotopes must be used. Other useful isotopes are $^{210}_{82}$Pb (half-life 22.3 years), $^{40}_{19}$K (half-life 1.28×10^9 years), and $^{238}_{92}$U (half-life 4.468×10^9 years).

29.4 Nuclear Stability and Binding Energy

Stable isotopes are found to occur naturally for all elements having proton numbers from 1 to 83 except for those with $Z = 43$ and 61. Although the nuclear interactions that explain nuclear stability are complicated, it is possible to obtain some general criteria of nuclear stability by looking at general features of stable nuclei.

Nucleon Populations
For a nucleus to be stable, the attractive strong nuclear force among its nucleons must be more than the repulsive force among the protons. Small nuclei are most stable when the number of protons and neutrons are approximately equal ($Z = N$). For example $^4_2$He, $^{12}_6$C, $^{23}_{11}$Na, and $^{27}_{13}$He have same or nearly the same number of protons and neutrons. Larger stable nuclei ($A > 40$) require more neutrons than protons to be stable. Heavy stable nuclei include $^{62}_{28}$Ni, $^{114}_{50}$Sn, $^{208}_{82}$Pb, and $^{209}_{83}$Bi. It is found that if the number of protons of a nucleus is more than 83, the strong force cannot overcome the repulsive force among the protons. Bismuth, $Z = 83$, is the stable nucleus with the maximum number of protons. Radioactive decay adjusts the number of protons and

2007 Pearson Education, Inc., Upper Saddle River, NJ. All rights reserved. This material is protected under all copyright laws as they currently exist. No portion of this material may be reproduced, in any form or by any means, without permission in writing from the publisher.

neutrons in an unstable nucleus until a stable nuclide is formed.

Pairing
Many stable nuclei have even numbers of both protons and neutrons, indicating that the protons and neutrons in a nucleus tend to pair up. Because of the **pairing effect** it was found that:
(i)   Most even-even nuclei are stable.
(ii)   Many odd-even or even-odd nuclei are stable.
(iii)   Only four odd-odd nuclei are stable
        ($^2_1 H$, $^6_3 Li$, $^{10}_5 B$, and $^{14}_7 N$).

Binding Energy
The masses of nuclei and nucleons are often expressed in atomic mass units, u, given by
$$1 \text{ u} = 1.66054 \times 10^{-27} \text{ kg}$$
The mass of a proton is 1.007276 u, and the mass of a neutron is 1.008664 u.

The atomic mass unit is equivalent to energy,
$$E = mc^2 = 931.5 \text{ MeV} \quad \text{where } 1 \text{ MeV} = 10^6 \text{ eV}.$$

The mass of a nucleus is found to be less than the total mass of its component nucleons. In general, for any nucleus, the **total binding energy** is related to the mass defect, Δm, by

$$E_b = \Delta mc^2 \tag{29.5}$$

For example, the helium atom has a mass of 4.002 603 u. The total mass of nucleons (two protons and two neutrons) in a helium nucleus is 4.032 980 u. Thus, a helium nucleus is less massive than the sum of its parts by $\Delta m = 0.030\ 366$ u. This mass difference is equivalent to an energy of

© 2007 Pearson Education, Inc., Upper Saddle River, NJ. All rights reserved. This material is protected under all copyright laws as they currently exist.
No portion of this material may be reproduced, in any form or by any means, without permission in writing from the publisher.

$(0.030377)(931.5 \text{ MeV/u}) = 28.30$ MeV. This energy is the total binding energy of the ^{4}He nucleus. This gives an average binding energy per nucleon 7.075 MeV for the helium nucleus.

The binding energy per nucleon increases rapidly with mass number A to a maximum at $A = 60$, then it decreases slowly and becomes nearly constant at about 7.2 MeV per nucleon for larger A. The nuclei in the range $A = 50$ to $A = 75$ are the most stable, and they have a binding energy per nucleon of about 8.3 MeV. Clearly, if a large unstable nucleus were split into two lighter stable nuclei, energy would be released. Similarly, if we could fuse together two light nuclei into a stable relatively heavier nucleus, energy would also be released.

Magic Numbers
In inert gases the outer shells of electrons are full and these gases are unreactive. An analogous effect is observed in the nucleus that indicates the existence of closed nuclear shells when the number of protons or neutrons is 2, 8, 20, 28, 0, 82, or 126. These numbers are called **magic numbers**. If a nucleus has a magic number of protons, it has an unusually high number of stable isotopes. The magic numbers are associated with unusually large binding energies, showing the high stability of the nucleus.

29.5 Radiation Detection and Applications

Detecting Radiation
A **radiation detector** detects radiation based on the particles produced by the radiation through interactions of the radiation with matter.

2007 Pearson Education, Inc., Upper Saddle River, NJ. All rights reserved. This material is protected under all copyright laws as they currently exist. No portion of this material may be reproduced, in any form or by any means, without permission in writing from the publisher.

In a *Geiger counter*, the incoming radiation ionizes a noble gas atom in the counter, thus freeing an electron that is accelerated toward the central positive electrode. On the way, the electron produces additional electrons through ionization of atoms of the gas. This results in a current pulse that is detected as a voltage across an external resistance.

In a *scintillation counter* a photon is emitted by a phosphor atom excited by an incoming particle. This causes the emission of a photoelectron from the photocathode of a *photomultiplier*. Accelerated through a potential difference in the photomultiplier, the photoelectrons emit secondary electrons as they collide with the successive electrodes at higher potentials. After several steps, weak scintillations are converted into sizable electrical pulses that are counted electronically.

In a *solid state* or *semiconductor detector* charged particles, as they pass through the semiconductor material, produce an electron-hole pair. The electron-hole pairs are collected to provide electric signals, which are then amplified and counted.

In some detectors the tracks of charged particles are recorded visually. In a *cloud chamber*, supercooled vapor condenses into droplets on ionized molecules created along the path of the particle. The path becomes visible as the droplets scatter light when illuminated. In a *bubble chamber* a reduction in pressure causes a liquid to be superheated and boil. Ions produced along the path of an energetic particle are sites for bubble formation and a trail of bubbles is created. A bubble chamber uses a liquid and the tracks are readily visible. In a *spark chamber* an incoming charged particle passes between a pair of

364

© 2007 Pearson Education, Inc., Upper Saddle River, NJ. All rights reserved. This material is protected under all copyright laws as they currently exist
No portion of this material may be reproduced, in any form or by any means, without permission in writing from the publisher.

electrodes having a high potential difference and immersed in a noble gas. The charged particles ionize the gas and show visible sparks between the electrodes that can be photographed.

Biological Effects and Medical Applications of Radiation

Nuclear radiation is harmful to living tissues. Alpha, beta, and gamma rays from radioactive decay can ionize thousands of atoms and molecules. As a result, a living cell exposed to such radiation may no longer function or behave normally, or it may soon die. In radiation treatments for cancer, high doses of radiation are used to kill the malignant cells of a cancerous tumor.

Radiation Dosage

The first radiation unit, the **roentgen** (R) is related to the amount of ionization caused by X-rays or gamma rays. The dosage of X-rays or gamma rays is 1 roentgen if an ionization charge of 2.58×10^{-4} C is produced in 1 kg of dry air.

The **rad** (*r*adiation *a*bsorbed *d*ose) is a measure of the amount of energy absorbed by an irradiated material, regardless of the type of radiation. If 1 kg of a sample absorbs 10^{-2} J of energy, it has received a dose of 1 rad. **Gray (Gy)** is the SI unit of absorbed dose,
$$Gy = 1 \text{ J/kg} = 100 \text{ rad.}$$

Relative biological effectiveness (RBE) takes into account that different types of radiation cause different amounts of damage. It is defined in terms of the number of rads of X-rays or gamma rays that produce the same biological damage as 1 rad of a given radiation.

2007 Pearson Education, Inc., Upper Saddle River, NJ. All rights reserved. This material is protected under all copyright laws as they currently exist. No portion of this material may be reproduced, in any form or by any means, without permission in writing from the publisher.

Combining the rad and RBE gives us the **rem** (*r*oentgen *e*quivalent in *m*an):

effective dose in rem = dose in rad × RBE   (29.6)

The SI unit of effective dose is **sievert (Sv)**.
Effective dose (in sieverts) = dose (in grays) × RBE (29.7)

Medical Treatment Using Radiation
Radioactive isotopes are used for medical treatment. For example, the radioactive isotope of iodine-131 is used to treat thyroid cancer.

Medical Diagnosis Applications That Use Radiation
Radioisotopes can be used for medical diagnostic procedures. For example, the activity of the thyroid gland can be determined by monitoring iodine uptake with small doses of iodine-123. Technetium-99 is used as a common tracer.

Gamma rays are also used for imaging purposes. In *single photon emission tomography* (SPET) a gamma detector is moved around the patient to measure gamma emission at different angles. *Positron-emission tomography* (PET) scans are used to locate tumors in the brain and also to find which regions of the brain are most active when a person performs a specific mental task. In a PET scan, a patient is given a radiopharmaceutical that decays by positron emission. The positrons encounter electrons in the body and they annihilate one another producing energetic gamma rays that move in opposite directions. Detectors determine the origin of the gamma rays and thus locate the areas where glucose metabolism is most intense.

© 2007 Pearson Education, Inc., Upper Saddle River, NJ. All rights reserved. This material is protected under all copyright laws as they currently exist.
No portion of this material may be reproduced, in any form or by any means, without permission in writing from the publisher.

Domestic and Industrial Applications of Radiation
In a common smoke detector at home, a weak radioactive
source ionizes the air and sets up a current. When the
smoke particles in the detector reduce the current, the
alarm sounds. In industry radioactive tracers are used to
determine flow rates in pipes, to detect leaks, and to study
corrosion and wear. **Neutron activation analysis** is used
for identification of bombs in airport luggage. In this case
an unstable neutron emitter is used to bombard the sample
containing nitrogen-14 (virtually all explosives contain
nitrogen) by neutrons. The excited nitrogen-15 decays
with the emission of gamma rays whose energy is
analyzed.

Hints and Suggestions for Solving Problems

1. Useful constants and conversions for this chapter are:
   $1 \text{ u} = 1.6606 \times 10^{-27} \text{ kg} = 931.5 \text{ MeV/c}^2$
   Mass of electron = $9.109390 \times 10^{-31} \text{ kg} = 0.511 \text{ MeV/c}^2$
   Mass of proton = $1.672623 \times 10^{-27} \text{ kg} = 938.28 \text{MeV/c}^2$
   Mass of neutron = $1.674929 \times 10^{-27} \text{ kg} = 939.57 \text{ MeV/c}^2$
   1 fermi (fm) = 10^{-15} m
   1 curie (Ci) = 3.7×10^{10} decays/s
   1 becquerel (Bq) = 1 decay/s
   1 roentgen (R) = 2.58×10^{-4} C/kg
   1 rad = 10^{-2} J/kg

2. Remember that $A = Z + N$. A nucleus X with atomic
   number Z and mass number A is written as $_Z^A X$.

3. The following information will be useful for
   determining the missing term in a nuclear decay for

367

2007 Pearson Education, Inc., Upper Saddle River, NJ. All rights reserved. This material is protected under all copyright laws as they currently exist.
No portion of this material may be reproduced, in any form or by any means, without permission in writing from the publisher.

some problems in Section 29.2. The total number of protons and neutrons is the same before and after an alpha decay. In a β^- decay, the number of protons is increased by one, and the number of neutrons is decreased by one. In a β^+ decay, the number of protons is decreased by one, and the number of neutrons is increased by one.

4. If N_0 is the number of nuclei at $t = 0$, the number of nuclei, N, remaining at time t is given by $N = N_0 e^{-\lambda t}$, where λ is the decay constant. From this equation, the half-life $t_{1/2}$ is $= 0.693/\lambda$. Use the preceding equation to determine the half-life from the decay constant, or vice versa. When calculating N, be sure to use the proper units of λ and t so that the product λt is dimensionless (that is, if t is in years, λ must be in 1/years). The preceding equations will be useful for solving problems in Section 29.3 including the problems on radioactive dating.

5. The nuclear binding energy (E_b) is the difference in mass between the nucleus and the total mass of its component nucleons times the square of the speed of light, $E_b = \Delta mc^2$. Use this relation to solve some problems in Section 29.4.

© 2007 Pearson Education, Inc., Upper Saddle River, NJ. All rights reserved. This material is protected under all copyright laws as they currently exist. No portion of this material may be reproduced, in any form or by any means, without permission in writing from the publisher.

CHAPTER 30
NUCLEAR REACTIONS AND
ELEMENTARY PARTICLES

This chapter discusses nuclear reactions including nuclear fission and nuclear fusion. Fundamental forces of nature, elementary particles, and, quark model are also discussed.

Key Terms and Concepts

Nuclear reactions
Nuclear fission
Liquid drop model
Chain reaction
Nuclear reactor
Nuclear fusion
Neutrino
Fundamental forces
Virtual particle
Mesons and pions
Weak nuclear force
Graviton
Elementary particles
Leptons, hadrons, and baryons
Quarks
Quark confinement
Electroweak force
Grand unification theory (GUT)

30.1 Nuclear Reactions

In **nuclear reactions** the original nuclei are converted into the nuclei of other elements. In 1919 Rutherford first induced a nuclear reaction, where a nitrogen nucleus was bombarded with alpha particles from bismuth-214. The

369

2007 Pearson Education, Inc., Upper Saddle River, NJ. All rights reserved. This material is protected under all copyright laws as they currently exist. No portion of this material may be reproduced, in any form or by any means, without permission in writing from the publisher.

nitrogen nucleus was *artificially transmuted* into an oxygen nucleus in the reaction:

$$^{14}_{7}\text{N} + {}^{4}_{2}\text{He} \rightarrow {}^{17}_{8}\text{O} + {}^{1}_{1}\text{H}$$

In a nuclear reaction like the preceding reaction a short-lived intermediate compound nucleus is formed in a highly excited state, as shown below:

$$^{14}_{7}\text{N} + {}^{4}_{2}\text{He} \rightarrow ({}^{18}_{9}\text{F}^*) \rightarrow {}^{17}_{8}\text{O} + {}^{1}_{1}\text{H}$$

Conservation of Mass–Energy and the Q Value

In a nuclear reaction, the total relativistic energy ($E = K + mc^2$) is conserved. The Q value of a nuclear reaction is defined as the change in kinetic energy, ΔK, in the reaction. For a reaction in which nitrogen is converted to oxygen,

$$Q = \Delta K = (K_O + K_P) - (K_N + K_\alpha) \qquad (30.1)$$

Q can be positive or negative. Thus, Q value is a measure of the energy released or absorbed in a nuclear reaction.

$$Q = (m_N + m_\alpha - m_O + m_p)c^2 \qquad (30.2)$$

For a general reaction $A + a \rightarrow B + b$, the Q value is given by:

$$Q = (m_A + m_a - m_B - m_b)c^2 = (\Delta m)c^2 \qquad (30.3)$$

When the Q value of a reaction is positive, the reaction is called **exoergic**. In this case, energy is released because some mass is converted into energy in the reaction. The Q value of a radioactive decay is always positive. If Q is negative, the reaction is called **endoergic** and energy must be given into the reaction to proceed. The minimum kinetic energy (K_{min}) that an incident particle must have to

© 2007 Pearson Education, Inc., Upper Saddle River, NJ. All rights reserved. This material is protected under all copyright laws as they currently exist.
No portion of this material may be reproduced, in any form or by any means, without permission in writing from the publisher.

initiate an endoergic reaction is called the **threshold energy**. For a nonrelativistic case, the threshold energy is:

$$K_{min} = \left(1 + \frac{m_a}{M_A}\right)|Q| \qquad (30.4)$$

where m_a and M_A are the masses of the incident particle and the stationary nucleus, respectively.

Reaction Cross Sections

Usually more than one reaction is possible when a particle collides with a nucleus. The probability of a particular nuclear reaction is given by the **cross section** of the reaction. The cross section depends on several factors including the kinetic energy of the incoming particle. For positively charged incoming particles, the reaction cross section generally increases with the kinetic energy of the incident particle. If the incident particle is a neutron (neutral), the cross section is found to vary with the neutron energy. The peaks where the cross sections are greatest are called **resonances**.

30.2 Nuclear Fission

The splitting of a heavy nucleus into two less-massive nuclei is called **nuclear fission**. A typical nuclear fission reaction generates two to three neutrons together with the daughter nuclei. Nuclear fission releases an amount of energy that is many orders of magnitude greater than the energy released in chemical reactions.

The following equation shows a typical fission reaction initiated by the absorption of a slow neutron by a uranium-235 nucleus.

371

2007 Pearson Education, Inc., Upper Saddle River, NJ. All rights reserved. This material is protected under all copyright laws as they currently exist. No portion of this material may be reproduced, in any form or by any means, without permission in writing from the publisher.

$$\,^{235}_{92}\text{U} + \,^{1}_{0}\text{n} \rightarrow \left(\,^{236}_{92}\text{U}^{*} \right) \rightarrow \,^{140}_{54}\text{Xe} + \,^{94}_{38}\text{Sr} + 2\,^{1}_{0}\text{n}$$

where $\,^{236}_{92}\text{U}^{*}$ is the intermediate unstable excited nucleus.

According to the *liquid drop model* the $\,^{236}_{92}\text{U}$ nucleus becomes distorted like a liquid drop and the repulsive electric force between the two parts causes the nucleus to split. Two other possible ways that fission of uranuim-235 can occur are:

$$\,^{235}_{92}\text{U} + \,^{1}_{0}\text{n} \rightarrow \,^{141}_{56}\text{Ba} + \,^{92}_{36}\text{Kr} + 3\,^{1}_{0}\text{n}$$

$$\,^{235}_{92}\text{U} + \,^{1}_{0}\text{n} \rightarrow \,^{150}_{60}\text{Nd} + \,^{81}_{32}\text{Ge} + 5\,^{1}_{0}\text{n}$$

When a heavy nucleus such as uranium splits, the average binding energy per nucleon increases from about 7.8 MeV to about 8.8 MeV. Thus, the energy liberated is about 1 MeV per nucleon, and the energy released per fission is about 234 MeV (since there are 234 nucleons in the uranium nucleus).

The neutrons released in one fission reaction may initiate additional fission reactions in other nuclei. A fission reaction that proceeds from one nucleus to another in this fashion is called a **chain reaction**. With more than 200 MeV of energy released per fission reaction, an uncontrolled chain reaction (in which more than one neutron initiates additional fission) can generate an incredible amount of energy in a very short amount of time. The minimum mass required to create a sustained chain reaction is known as the **critical mass**. Natural uranium contains 0.7% of ^{235}U and 99.3% of ^{238}U. Since ^{238}U may absorb neutrons and trigger nuclear reactions other than nuclear fission, the ^{235}U concentration in the sample is increased to have more fissionable nuclei. In an

© 2007 Pearson Education, Inc., Upper Saddle River, NJ. All rights reserved. This material is protected under all copyright laws as they currently exist.
No portion of this material may be reproduced, in any form or by any means, without permission in writing from the publisher.

atom bomb an uncontrolled chain reaction produces an enormous amount of energy almost instantaneously. By limiting the number of neutrons in the environment of the fissile nuclei, it is possible to release energy from nuclear fission in a controlled manner. This is done in a **nuclear reactor**.

Nuclear Reactors
The Power Reactors
The five key elements to a reactor are: fuel rods, core, coolant, control rods, and moderator.

Fuel rods are tubes packed with pellets of uranium oxide placed at the central portion of the reactor, called the **core**. **Coolant** such as water is used to remove heat from the chain reaction. Heavy water is also used as a coolant with the advantage that deuterium does not readily absorb neutrons. Water also acts as a **moderator** to slow down the speed of neutrons by collisions. **Control rods** (boron or cadmium) are inserted into or withdrawn from the reactor core to control the chain reaction rate and, hence, the released energy. In practice, energy production is monitored electronically and the operation is stabilized by a *feedback* mechanism.

The Breeder Reactor
^{238}U has an appreciable fission cross section for *fast* neutrons. In a **breeder reactor**, by reduced moderation one or more neutrons from ^{235}U fission are absorbed by ^{238}U nuclei to produce ^{239}Pu. ^{239}Pu is fissionable and serves as additional fuel.

Electricity Generation
The generated heat in a controlled chain reaction in a nuclear reactor is carried away by water surrounding the rods in the fuel assembly. The hot water is pumped to a

2007 Pearson Education, Inc., Upper Saddle River, NJ. All rights reserved. This material is protected under all copyright laws as they currently exist. No portion of this material may be reproduced, in any form or by any means, without permission in writing from the publisher.

heat exchanger to transfer energy to the water of a steam generator. High-pressure steam turns a turbine to generate electric energy.

Nuclear Reactor Safety

Since the coolant also serves as a moderator, the chain reaction would stop if the coolant is lost. In such as case, the fuel rods may become hot enough for the cladding to melt and fracture. In this case, the hot fissioning mass could fall into the water and cause a steam explosion, a hydrogen explosion, or both. This also may rupture the walls of the vessel and cause radioactive fragments to come into the environment. Also, the fission fragments are radioactive and have long half-lives. Thus, radioactive waste is a problem even when reactors operate safely.

30.3 Nuclear Fusion

When two light nuclei combine to form a relatively massive nucleus, the process is called **nuclear fusion**. In nuclear fusion, the larger nucleus formed in the fusion reaction has less mass than the sum of the masses of the original less massive nuclei. The mass difference appears as the liberated energy (according to $E = mc^2$) in the fusion.

For example, the fusion of deuterium and tritium to form helium is:

$$\ {}_{1}^{2}\text{H} \quad + \quad {}_{1}^{3}\text{H} \quad \rightarrow \quad {}_{2}^{4}\text{He} \quad + \quad {}_{0}^{1}\text{n}$$

$$\text{(2.014 102 u)} \quad \text{(3.016 049 u)} \quad \text{(4.002 603 u)} \quad \text{(1.008 665 u)}$$

This fusion reaction liberates 17.6 MeV per fusion.

© 2007 Pearson Education, Inc., Upper Saddle River, NJ. All rights reserved. This material is protected under all copyright laws as they currently exist.
No portion of this material may be reproduced, in any form or by any means, without permission in writing from the publisher.

Nuclear fusion is the source of energy of stars including the sun. Most stars, including the sun, fuse hydrogen to produce helium. The sun is powered by the *proton–proton cycle* of fusion reactions, in which fusion is completed in three steps. The net effect is that four protons combine to form a helium nucleus plus two positrons, two gamma rays (γ), two neutrinos (ν), and energy release, as shown below:

$$4\,({}^{1}_{1}\text{H}) \;\longrightarrow\; {}^{4}_{2}\text{He} + 2\,({}^{0}_{+1}\text{e}) + 2\gamma + 2\nu + Q$$
$$(Q = +24.7 \text{ MeV})$$

In the sun fusion involves only 10% of the solar mass. It has been going on for about 5 billion years and is expected to continue for about another 5 billion years. After that, nuclear fusion of helium will create heavier elements such as lithium and carbon.

Fusion as a Source of Energy Produced on the Earth
Fusion, such as that of deuterium present in the oceans, is an ideal source of energy of the future. There is less danger of releasing radioactive material in the reaction than in nuclear fission. However, to start nuclear fusion, the initial lighter nuclei must have enough energy to overcome their mutual Coulomb repulsion. The needed temperature is about 10^{7} K to initiate *thermonuclear fusion reactions*. At such high temperatures neutral atoms are broken into a state, called **plasma**, containing positively charged ions and negatively charged electrons. Since plasma consists of charged particles, electric and magnetic fields can control it. In **magnetic confinement**, magnetic fields hold plasma in a confined space and electric fields are used to produce electric currents in it that raise its temperature to approximately 100 million kelvins. In **inertial confinement** hydrogen fuel pellets are dropped or

375

2007 Pearson Education, Inc., Upper Saddle River, NJ. All rights reserved. This material is protected under all copyright laws as they currently exist. No portion of this material may be reproduced, in any form or by any means, without permission in writing from the publisher.

positioned in a reactor chamber. Pulses of laser or electron or ion beams implode the pellets to produce compression, high densities, and high temperatures.

30.4 Beta Decay and the Neutrino

Two basic β^- and β^+ decays are, respectively:

$$^{14}_{6}C \quad \longrightarrow \quad ^{14}_{7}N + ^{0}_{-1}e$$

$$^{13}_{7}N \quad \longrightarrow \quad ^{13}_{6}C + ^{0}_{+1}e$$

When the preceding β^- decay is analyzed, the energy released or the Q value (that comes from the mass difference in the reaction) is found to be about 0.156 MeV. However, most electrons in the decay that would carry this energy are observed to have energy less than this value. The β^- decay thus violates the conservation of energy. This decay also violates the conservation of momentum since the parent nucleus, daughter nucleus, and the emitted electron do not always move in opposite directions. The total angular momentum also is not found to be same before and after the decay.

Pauli resolved these apparent difficulties by proposing that in addition to an electron another unobserved particle, called a **neutrino**, is created in the β^- decay. A neutrino has zero mass, intrinsic spin ½, and it travels with the speed of light. It was found later that a neutrino interacts with matter (*weak interaction*) through a force much weaker than the strong force, and that is why it was not

© 2007 Pearson Education, Inc., Upper Saddle River, NJ. All rights reserved. This material is protected under all copyright laws as they currently exist No portion of this material may be reproduced, in any form or by any means, without permission in writing from the publisher.

observed in the decay. The neutrino was observed
experimentally in 1956. The correct decay equations are:

$$^{14}_{6}C \quad \rightarrow \quad ^{14}_{7}N + ^{0}_{-1}e + \bar{\nu}_e$$

$$^{13}_{7}N \quad \rightarrow \quad ^{13}_{6}C + ^{0}_{+1}e + \nu_e$$

ν_e represents a neutrino and the one with a bar over it
represents an *antineutrino*. The antineutrino is associated
with the decay of a free neutron whereas the neutrino is
associated with the decay of a proton.

30.5 Fundamental Forces and Exchange Particles

There are four **fundamental forces** in nature. In order of
diminishing strength, these are the *strong nuclear force,
the electromagnetic force, the weak nuclear force, and the
force of gravity.*

Gravity and electromagnetic forces obey an inverse square
relationship. Classically, these forces use the concept of
field. Quantum mechanically, forces are transmitted via
the exchange of particles. For *extremely* short time
intervals creation of such particles is permitted by the
uncertainty principle and energy conservation is briefly
violated. The created particle is absorbed before it is ever
detected, and that is why it is called a **virtual particle**.
Thus, fundamental forces are carried by virtual **exchange
particles**. *The range of a force associated with an
exchange particle is inversely proportional to the mass of
that exchange particle.*

©2007 Pearson Education, Inc., Upper Saddle River, NJ. All rights reserved. This material is protected under all copyright laws as they currently exist.
No portion of this material may be reproduced, in any form or by any means, without permission in writing from the publisher.

The Electromagnetic Force and the Photon

A (virtual) **photon** is the exchange particle of the electromagnetic force. The photon is a massless particle, and that is why the electromagnetic force is of infinite range. The *Feynman diagram* visualizes a particle exchange such as for repulsion between two electrons.

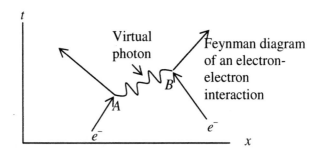

The interacting electrons undergo a change in energy and momentum due to exchange of a virtual photon which is created at a point A and absorbed at another point B in an amount of time that is consistent with the uncertainty principle.

The Strong Nuclear Force and Mesons

An exchange particle, called a **meson**, is associated with the strong nuclear force between two nucleons. Using the uncertainty principle, the mass of a meson (m_m) was estimated as $m_m \approx 270\ m_e$, m_e being the mass of an electron.

© 2007 Pearson Education, Inc., Upper Saddle River, NJ. All rights reserved. This material is protected under all copyright laws as they currently exist. No portion of this material may be reproduced, in any form or by any means, without permission in writing from the publisher.

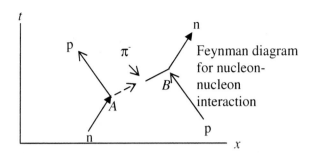

Feynman diagram for nucleon-nucleon interaction

Although a virtual meson cannot be detected, real mesons can be detected if sufficient energy is involved in the collision of nucleons. In 1936, the **muon** (also called mu meson) was discovered in cosmic rays with two charge varieties, $\pm e$ and of mass $m_{\mu\pm} = 207\ m_e$. But, the muon does not behave like a strong interacting particle. In 1947, new particles, called **pions**, were discovered in cosmic rays that interact strongly with matter and are believed to be the exchange particle for the strong nuclear force. The masses of pions (one positive, one negative, and one neutral) are:

$$m_{\pi\pm} = 273\ m_e \text{ and } m_{\pi 0} = 264\ m_e$$

Free pions and muons are unstable. They decay as:

$$\pi^+ \rightarrow \mu^+ + \nu_\mu \text{ (muon neutrino)}$$

$$\mu^+ \rightarrow \beta^+ + \nu_e + \overline{\nu}_\mu$$

The Weak Nuclear Force and the W Particle

Experiments show that free neutron decay can be expressed as:

$$n \rightarrow p^+ + e^- + \overline{\nu}_e$$

with a half-life of 10.4 min. The decay rate measurements show that the force that causes a neutron to disintegrate is

379

©2007 Pearson Education, Inc., Upper Saddle River, NJ. All rights reserved. This material is protected under all copyright laws as they currently exist. No portion of this material may be reproduced, in any form or by any means, without permission in writing from the publisher.

weaker than electromagnetic but stronger than gravity. It is called the **weak nuclear force**. The weak force has a range of about 10^{-17} m and the virtual exchange particles of the weak nuclear force are called W particles. W particles have masses about 100 times the mass of a proton. That explains why the force is of extremely short range. High-energy accelerators are able to create real W particles, confirming the existence of W particles.

The Gravitational Force and Graviton

The exchange particle of the gravitational force is called the **graviton**. The graviton is massless and has not yet been observed because of the relative weakness of its interaction.

30.6 Elementary Particles

The elementary particles are the fundamental building blocks of all atoms. Among the three subatomic particles (electrons, protons, and neutrons), only the electron is considered to be elementary, whereas protons and neutrons are composed of smaller elementary particles. About 300 new particles have been discovered so far, most of which are unstable.

Leptons

The elementary particles that experience the weak nuclear force are called leptons. There are only six leptons, including the electron, muons, and, the **tauon**. Since no internal structure has been detected in any of these leptons, they are considered to be elementary particles. Muons are electrically charged and about 200 times more massive than an electron. The tauon has a mass twice that of a proton. The electron, muon, and tauon are negatively charged and have positively charged antiparticles.

380

© 2007 Pearson Education, Inc., Upper Saddle River, NJ. All rights reserved. This material is protected under all copyright laws as they currently exist. No portion of this material may be reproduced, in any form or by any means, without permission in writing from the publisher.

Neutrinos are also leptons. There are three types of neutrinos [called the electron neutrino (ν_e), the muon neutrino (ν_μ), and the tau neutrino (ν_τ)], each associated with a different charged lepton ($e^\pm$, $\mu^\pm$, $\tau^\pm$).

Hadrons
Hadrons are composite particles that experience both the weak and the strong nuclear force. Hundreds of hadrons are known to exist, including the proton and the neutron. Hadrons are subdivided into *mesons* and *baryons*. **Baryons** include the proton and the neutron. Baryons have spin ½ or 3/2. **Mesons**, which include pions, have spins 0 or 1 and they eventually decay into leptons and photons.

30.7 The Quark Model

All hadrons are composed of a number of truly elementary particles called **quarks**. Since some hadrons are electrically charged, quarks also possess charge. According to Gell-Mann and Zweig there are three types of quarks. They have been named *up* (u), *down* (d), and *strange* (s) and their antiparticles (antiquark). *Quarks are the fundamental particles of the hadron family*. All quarks have charges that are fractions of the charge of the electron, *e*. The charge of u, d, and s quarks are $+(2/3)e$, $-(1/3)e$ and $-(1/3)e$, respectively. Anitquarks have charges opposite to those of quarks.

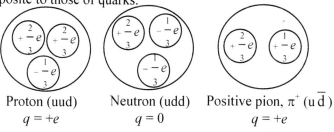

Proton (uud)	Neutron (udd)	Positive pion, π^+ (u $\bar{\text{d}}$)
$q = +e$	$q = 0$	$q = +e$

381

2007 Pearson Education, Inc., Upper Saddle River, NJ. All rights reserved. This material is protected under all copyright laws as they currently exist. No portion of this material may be reproduced, in any form or by any means, without permission in writing from the publisher.

For example, the proton is composed of two up quarks and one down quark (*uud*). The neutron is composed of one up quark and two down quarks (*udd*). Mesons are composed of bound pairs of quarks and antiquarks.

Quark Confinement, Color Charge, and Gluons

No free quark has been observed. It is believed that a free, independent quark cannot exist. The force between two quarks increases with the separation between them, and as a result, quarks must always be bound with other quarks. This concept is known as **quark confinement**.

Quarks are endowed with a characteristic, called **color charge** or simply color. Quarks of different colors attract each other via a force, called **color force**, that keeps them confined. The color force between two quarks occurs due to the exchange of virtual particles called **gluons**.

30.8 Force Unification Theories, the Standard Model, and the Early Universe

Unification Theories

The theories that attempt to unify various fundamental forces are known as **unification theories**. Glashow, Salam, and Weinberg (1960) combined the weak nuclear force with the electromagnetic force into a single **electroweak force**. According to the theory, weakly interacting particles such as electrons and neutrinos carry a *weak charge*. This weak charge is responsible for the exchange of particles, creating the combined electroweak force. The theory predicted three electroweak exchange particles, W^+ and W^-, when there is a charge exchange and the neutral Z° when there is no charge exchange. These particles have been discovered. Physicists are looking for

© 2007 Pearson Education, Inc., Upper Saddle River, NJ. All rights reserved. This material is protected under all copyright laws as they currently exist. No portion of this material may be reproduced, in any form or by any means, without permission in writing from the publisher.

the **grand unified theory** (GUT) that combines the electroweak force and the strong nuclear force. Final unification would be to combine the gravitational force with GUT creating a single **superforce**.

The Standard Model

The electroweak theory and the QCD model together is called the **standard model**. According to the model the gluons carry the strong force, which keeps quarks together to form composite particles such as protons and pions. Leptons do not experience the strong force and participate only in the gravitational and electroweak interactions. Gravitons presumably carry the gravitational interactions and photons carry the electroweak interaction.

Evolution of the Universe and the Superforce

The universe began from a huge explosion called *Big Bang* about fifteen billion years ago. The temperature during the first 10^{-45} s of the explosion was about 10^{32} K. At this time the four fundamental forces were combined into a single force. As the universe expanded and cooled the elementary particles that were created condensed to first form protons, neutrons, and electrons. In turn, they combined to form atoms and molecules, then stars and galaxies. Over billions of years the universe cooled to the present temperature of about 3 K. In this process the superforce symmetry was lost leaving four different-looking fundamental forces.

2007 Pearson Education, Inc., Upper Saddle River, NJ. All rights reserved. This material is protected under all copyright laws as they currently exist. No portion of this material may be reproduced, in any form or by any means, without permission in writing from the publisher.

Hints and Suggestions for Solving Problems

1. Useful constants and conversions for this chapter are:
   $1 \text{ u} = 1.6606 \times 10^{-27}$ kg $= 931.5$ MeV/c^2
   Mass of electron $= 9.109390 \times 10^{-31}$ kg $= 0.511$ MeV/c^2
   Mass of proton $= 1.672623 \times 1^{-27}$ kg $= 938.28$ MeV/c^2
   Mass of neutron $= 1.674929 \times 10^{-27}$ kg $= 939.57$ MeV/c^2

2. To determine the energy released in a nuclear fission or nuclear fusion determine the mass difference before and after the reaction, then use Einstein's relation $E = |\Delta m|c^2$. Usually, the energy released is expressed in MeV. Use the necessary conversion factor to convert the energy to MeV. Note that for energy to be released in a nuclear reaction, the mass difference (final mass minus initial mass) has to be negative.

© 2007 Pearson Education, Inc., Upper Saddle River, NJ. All rights reserved. This material is protected under all copyright laws as they currently exist. No portion of this material may be reproduced, in any form or by any means, without permission in writing from the publisher.